Environmental Biotechnology and Biodiversity Conservation

About the Authors

Dr. Mihir Kumar Das is a Senior Lecturer in Post-Graduate Dept. of Botany, Gangadhar Meher College (Autonomous), Sambalpur. He did his Ph.D. from Utkal University in 1997. He has been imparting teaching in taxonomy, microbiology, biotechnology etc. at under-graduate, post-graduate and M.Phil. levels. He has significant contribution in the field of cyanobacterial diversity and also on Angiospermic taxonomy. He has attended many national and international conferences and workshops. He has published many research articles in reputed journals. He has completed two U.G.C. research projects. He has guided many M.Phil. Scholars. He writes popular articles in the science magazines, newspapers and also delivers talk on AIR, Doordarshan for diffusion of scientific knowledge. He is the life member of many-learned societies. He was conferred an honorary appointment to the Research Board of Advisors since 2002 by the American Biographical Institute, Inc. and also the award of " Scientist of the Year-2004" by National Environmental Science Academy (NESA), New Delhi.

Environmental Biotechnology and Biodiversity Conservation

Edited By
Dr. Mihir Kumar Das
Senior Lecturer in Botany
G.M. College (Autonomous), Sambalpur (Orissa)

2019
Daya Publishing House®
A Division of
Astral International Pvt. Ltd.
New Delhi – 110 002

First Published: 2008

Reprinted 2019

ISBN: 9789351240242

Publisher's Note:

Every possible effort has been made to ensure that the information contained in this book is accurate at the time of going to press, and the publisher and author cannot accept responsibility for any errors or omissions, however caused. No responsibility for loss or damage occasioned to any person acting, or refraining from action, as a result of the material in this publication can be accepted by the editor, the publisher or the author. The Publisher is not associated with any product or vendor mentioned in the book. The contents of this work are intended to further general scientific research, understanding and discussion only. Readers should consult with a specialist where appropriate.

Every effort has been made to trace the owners of copyright material used in this book, if any. The author and the publisher will be grateful for any omission brought to their notice for acknowledgement in the future editions of the book.

Published by : Associated Publishing Company®
A Division of
Astral International Pvt. Ltd.
– ISO 9001:2015 Certified Company –
4736/23, Ansari Road, Darya Ganj
New Delhi-110 002
Ph. 011-43549197, 23278134
E-mail: info@astralint.com
Website: www.astralint.com

Preface

Biotechnology has made profound impact in the field of health, food, agriculture and environmental protection. It has the potential to provide a wide array of benefits to humanity, including treatment for hither to incurable diseases, safer, cheaper and more effective drugs, more nutritional diary and agricultural products, more resilient and productive crop and tree species, new sources of renewable energy and more effective treatment of wastes. India with its large pool of scientific talent, world class IT industry and vibrant pharmaceutical sector is well positioned to emerge as significant player in the global biotech area. The Government of India realized the potential and benefits of the industry at an early stage when the biotech wave was at best, just a ripple and has initiated several reforms and measures for the development of the industry. The combined efforts of Central and State Governments, industries and the scientific communities have resulted in patenting of innovations, technology transfer to industries and a new direction to biotechnology research in India.

Environmental biotechnology is concerned both with the implications and applications of biotechnology in the wider context of environment. Due to rapid industrialization, urbanization and other developments, there is a constant threat to the clean environment and to the depleting natural resources. The first conference on the Human environment, held at Stockholm in 1972;

the United Nations Conference on Environment and Development (UNCED) at Rio de Janeiro in 1992 and the latest conference the World Summit on Sustainable Environment (WSSE) at Johannesburg in 2002 bear the testimony of global concern of environmentalist. While on the one hand, there is an increasing problem of control of environmental pollution; there is also a problem of the conservation of nature and natural resources. Both these problems are receiving constant attention of environmentalist. Among implications, there is also an alarm due to release of genetically engineered organisms in the atmosphere and due to release of effluents from biotechnological companies, so that the environmentalist are having a debate on the effects of developments in biotechnology on the environment. On the other hand, efforts are also being made to use biotechnology to protect the environment from pollution and to conserve natural resources. The environment includes various ecological sectors, which can be divided into three major categories: forest ecosystems including reserved forest, coastal ecosystems harboring the mangrove vegetation and many inland sectors having mining activities. The ***ex situ*** and ***in situ*** conservation of forest species should also be supplemented with molecular taxonomic approaches.

The rapid advances and developments in the field of environmental biotechnology and biodiversity conservation ignited in my mind to gather latest research works in the form of a book, which can fulfill the needs of the Post-graduate students as well as the researchers. This book consists of 16 chapters, encompassing topics on various fields like biotechnological applications of algae, ***in vitro*** propagation of plant, stress responses, environmental impact of pesticides, biodiversity study, ethnobotany and many others. Besides a few review articles, almost all chapters deal with the experimental findings, which will enhance the scientific temper among the students, as well as, provide clues to for future research work. I hope, this book would be informative to the students, teachers, scientists and researchers in the field of applied sciences.

The author expresses his sincere thanks to the contributors, whose articles have enriched this volume. I express my deep sense of gratitude to Prof. Shiba P. Adhikary, a renowned scientist in the field of algology, who has inspired me to compile these works in the form of a book. Last but not the least, I am extremely thankful to

Mr. Anil Mittal, publisher of Daya Publishing House, New Delhi for undertaking the publication of this book. Finally, I am indebted to all our well wishers for their blessings without whom this book could not have been completed.

Mihir Kumar Das

Contents

List of Contributors

A.K. Bastia

P.G. Department of Botany, North Orissa University, Baripada–757 003, Orissa

Amarendra Mishra

Banki Autonomous College, Banki, Orissa

B.B. Mishra

Minor Irrigation and Water Use Training Institute, Bhubaneswar, Orissa

B.L. Bohidar

Department of Botany, Women's College, Jharsuguda–768 201, India

B.N. Mahapatra

Gopabandhu Ayurved Mahavidyalaya, Puri – 750 002, Orissa

B.P. Choudhury

P.G. Department of Botany, Utkal University, Bhubaneswar

D. Panda

Division of Biochemistry, Plant Physiology and Environmental Sciences, Central Rice Research Institute, Cuttack–753 006, Orissa

Geeta Rani Pattanaik

P.G. Department of Botany, G.M.College (Autonomous), Sambalpur–768 004

H. Dey

P.G. Department of Botany, North Orissa University, Baripada–757 003, Orissa

J.K. Bandyopadhyay

Experimental Fish Farm and Field Center NRC CWF, Collectorate Road, Champawat, Uttaranchal–262 523

Keshab K. Mohapatra

Chief District Veterinary Office, Sambalpur

Khageswar Pradhan

Department of Environmental Science, Jyoti Vihar, Burla–768 019, Orissa

Kunja B. Satapathy

Department of Botany, Vyasanagar College, Jajpur Road–755 019, Orissa

L.M. Behera

P.G. Department of Botany, G.M. College (Autonomous), Sambalpur–768 004, Orissa

Mihir K. Das

P.G. Department of Botany, G.M. College (Autonomous), Sambalpur–768 004, Orissa

M.M. Prasad

Central Institute of Fisheries Technology, Burla Research Centre, Burla–768 017, Sambalpur, Orissa

P.K. Chand

Post-Graduate Department of Botany, Utkal University, Vani Vihar, Bhubaneswar–751 004, Orissa

Prem Kumar

Experimental Fish Farm and Field Centre, NRC CWF, Collectorate Road, Champawat, Uttaranchal–262 523.

R.K. Nayak

Department of Botany, G.M. College, Sambalpur–768 004, India

R.K. Sarkar

Division of Biochemistry, Plant Physiology and Environmental Sciences Central Rice Research Institute, Cuttack–753 006, Orissa, India

Rajan Kumar Gupta

Department of Botany, Government Post-Graduate College, Rishikesh–249 201, Dehradun, Uttarakhand

S. Bhakta

P.G. Department of Botany, North Orissa University, Baripada–757 003, Orissa

S.G. Sharma

Division of Biochemistry, Plant Physiology and Environmental Sciences, Central Rice Research Institute, Cuttack–753 006, Orissa, India

S.L. Swain

P.G. Department of Botany, G.M.College (Autonomous), Sambalpur–768 004, India

Shiba P. Adhikary

P.G. Department of Botany, Utkal University, Bhubaneswar–751 004.

Uma Devi

Department of Botany, Government Women's College, Sambalpur–768 001, Orissa

Vinod Kumar

Indian Grassland and Forage Research Institute, Jhansi

Chapter 1

Biotechnological Applications of Algae

Rajan Kumar Gupta*

Department of Botany, Government Post Graduate College (Identified as "College with potential for Excellence by UGC and NAAC Accredited "A" Grade College Rishikesh – 249 201 (Dehradun), Uttarakhand

ABSTRACT

The algae are (with numerous exceptions) aquatic organisms that (with frequent exceptions) are photosynthetic, oxygenic autotrophs that are (except for the kelps) typically smaller and less structurally complex than land plants. However, the blue-green algae are prokaryotic and the most simplest of the algae. Blue-green algae also known as cyanobacteria are the first recognized species on earth, they are blue-green in color and hence the name blue-green has been given to them. They have been living on earth for over 2.5 billion years since Pre-Cambrian era. They occupy a variety of ecological niches ranging from Arctic to Antarctic from freshwater to saline and psychrophilic, thermophilic, acidophilic and alkalophilic forms have also been reported. They are a group of gram negative photosynthetic prokaryotes which possess oxygen evolving photosynthetic apparatus similar to higher plant chloroplasts.

* E-mail: rajankgupta1@rediffmail.com; Tel: 0135-2450373; Mob. 9412319164

A variety of them are able to fix atmospheric nitrogen. In the recent times cyanobacteria received a greater attention owing to their importance as food, feed, colorant, vitamins, toxins, enzymes, pharmaceuticals etc. The article mainly focus on the potential biotechnological applications of algae and cyanobacteria.

Keywords: *Biotechnology, Biofertilizer, Cyanobacteria, Neutraceuticals, Pharmaceuticals.*

Introduction

Morphologically diverse group of autotrophs, algae, are found to occupy almost every possible ecological niche on earth. They possess a photosynthetic apparatus similar to that of higher plant chloroplasts. During the course of evolution, these organisms were the first to derive electrons for CO_2 fixation from water, evolving O_2 as a by-product and were instrumental in changing the prevalent reducing conditions to an oxidizing one (Haselkorn, 1978; Wolk, 1996). In a way, the cyanobacteria are credited with providing an oxidizing environment for the development of many other microbes and higher eukaryotic life forms with complex metabolic pathways. Many of these forms are able to convert molecular nitrogen to ammonia with the help of the nitrogen fixing enzyme nitrogenase which is present in a highly specialized cell known as heterocyst. Some cyanobacteria, although do not form heterocysts fix atmospheric nitrogen efficiently (Stal, 2000). A survey of rice fields has shown that the population of non-heterocystous cyanobacteria is much higher than that of the heterocystous forms (Desikachary, 1959; Anand, 1989). According to an estimate, cyanobacteria constitute about 15 per cent of the total algal flora in the tropics about 2 per cent in the temperate regions (Thajuddinn and Subramanian, 2005).

Some cyanobacteria are of nuisance value as they grow luxuriantly and pollute a variety of water bodies by forming water-blooms. *Microcystis*, an abundantly occurring nuisance cyanobacterium in many eutrophic ponds and reservoirs of India and other tropical countries, is responsible for unpleasant odor of water bodies (Rai *et al.*, 1998). The cyanobacteria are important primary producers in many ecosystems including unfriendly extreme environments. Some of the cyanobacterial strains are extremophiles and are able to grow in extreme habitats with low and high pH, brines, low and high temperature, metal polluted soils and in environments where no other microalgae can exist (Broady *et al.*, 1996; Grant and Tindall, 1986; Gaur and Rai, 2001; Thajuddin *et al.*, 2002).

Cyanobacteria also form symbiotic associations with plants. Compatible cyanobacteria are recognized by respective host plants which influence their behavior. Symbiotic relationship exists with, for example bryophytes, fungi, pteridophytes, gymnosperms and angiosperms (Rai, 1990). In addition to this, they are potential source of food and medicine which make them potential organisms useful to mankind.

Commercial Applications

Much interest is currently being devoted to bioprospecting of cyanobacteria owing to the potential economic implications. Cyanobacteria offer immense potential as biofertilizer, food, feed, medicine etc. Since cyanobacteria contain a variety of fine chemicals such as pigments, vitamins and enzymes they have been widely used in scale-up technologies for the production of such chemicals. National centre for conservation and utilization of blue green algae (NCCUBGA) at Indian Agriculture Research Institute, New Delhi maintain collection of several freshwater cyanobacterial forms and is a prestigious National facility. Similarly a National facility for marine cyanobacteria (NFMC) was established at the Department of Microbiology, Bharatidasan University, Tiruchirapalli, and Tamilnadu. Many private companies have now established Research and Development wing focusing on the commercial exploitation of cyanobacteria. For example M/S ABL Biotechnologies, Chennai has a very active R&D set up for the commercial exploitation of cyanobacteria.

Source of Bioenergy

Among the algae microalgae are the most primitive form of plants. While the mechanism of photosynthesis in microalgae is similar to that of higher plants, they are generally more efficient converters of solar energy because of their simple cellular structure. In addition, because the cells grow in aqueous suspension, they have more efficient access to water, CO_2, and other nutrients. For these reasons, microalgae are capable of producing 30 times the amount oil per unit area of land, compared to terrestrial oilseed crops.

An extensive study has been made by many researchers on the microalgae which can be used as a source of biofuels like methane gas, alcohol or hydrocarbons or vegetable oils (Benemann *et al.*, 1986; Cohen, 1986). In early 1970's, the US department of energy has spent 10 million US dollars towards developing low cost technology using microalgae for fuel production, through its solar energy research institute aquatic species programe (Neenan *et al.*,1986). The production of various kinds of biofuel was reported in different organisms. Cultures of *Anabaena, Spirulina, Oscilatoria, Chlorogloea, Nostoc, Mastigocladus* and *Chlamydomonas reinharditii* were reported for the production of hydrogen (Hallen beck *et al.*, 1978; Mitsui, 1980; Kumazawa and Mitsui, 1982; Karube *et al.*, 1986; Lindblad *et al.*, 1998; Hall, 1998; Robinson *et al.*, 1986). In 1977 Keenan reported the production of methane gas by *Anabaena flosaquae.* Generation of electricity by immobilized cyanobacterial species *Mastigocladus* and *Phormidium* onto SnO_2 optically transparent electrodes (Robinson *et al.*, 1986). The alga *Botryococcus braunii* has been shown to yield 70 per cent of its extract is hydrocarbon liquid, which closely resembles crude oil.

In the late 1970s, the emphasis switched from methane production to algal oils as the fuel product. This was based on the known ability of some microalgae species to accumulate large amounts of algal lipids, in particular under conditions of nutrient (mainly N and Si) limitations. As before, 40-ha earthwork ponds were used; however, this time with paddle wheel mixing. Productivities were now projected of 67.5 mt/ha/yr for an algal biomass containing 40 per cent lipids (oils) by weight. This corresponded to about 90 mt/ha/yr for conventional algal biomass; yielding almost 160 barrels of crude oil/ha/yr. Harvesting was again assumed to be by bioflocculation, followed by a centrifugation process to concentrate the biomass to a paste-like consistency. A solvent extraction process as used for soybean oil extraction was assumed, at three-time higher unit cost to account for the high moisture in the paste. (However, as was pointed out, direct solvent extraction was unlikely to be feasible for such high moisture biomass). As before, the residual biomass was to be anaerobically digested in covered ponds to produce methane gas, with the nutrients (and C) from the digester (and digester gas) recycled to the ponds (Benemann *et al.*, 1982a). Four major research needs to achieve the objectives of high productivity in large-scale outdoor systems: 1. Photosynthetic efficiency for light energy and high lipid production. 2. Fundamentals of species selection and control in open pond systems. 3. Mechanisms (and control) of algal bioflocculation. 4. Effects of non-steady-state operating conditions on algal metabolism (Benemann *et al.*, 1982b).

Biofuel Production

Microalgae are microscopic organisms that can grow via photosynthesis. Many microalgae grow quite rapidly, and are considerably more productive than land plants and macroalgae (seaweed). Microalgae reproduce primarily by vegetative (asexual) cell division, although sexual reproduction can occur in many species under appropriate growth conditions. There are several main groups of microalgae, which differ in pigment composition, biochemical constituents, ultrastructure, and life cycle. Five groups were of primary importance for biofuel productions are: diatoms (Class: Bacillariophyceae), green algae (Class: Chlorophyceae), golden brown algae (Class: Chrysophyceae), prymnesiophytes (Class: Prymnesiophyceae), and the eustigmatophytes (Class: Eustigmatophyceae). The blue-green algae or cyanobacteria (Class: Cyanophyceae).

Hydrogen Production

Strains of *Synecococcus, Microcystis* and *Oscillatoria* have received much attention as a clean type of fuel. They produce hydrogen at a significant rate. The strains of Miami BG7 of *Oscillatoria* is particularly suitable as it lacks the uptake hydrogenase which catalyses the uptake of hydrogen by the cells. This enzyme has an adverse effect on the net yield of hydrogen gas. Philips and Mitsui observed that immobilization of the strain Miami BG7 with agar significantly improves the rates and duration of hydrogen production as compared to that by free cell suspensions. The rates of hydrogen production of about 12 microlitres H_2 per mg dry weight per hour were recorded and hydrogen production could be sustained for three weeks (Prescott, 1969; Kumar and Singh, 1995).

Hydrocarbon Production

For the estimation of hydrocarbons first of all total lipids were extracted with chloroform-methanol-water (5:10:4 by volume) according to the methods of Bligh and Dyer (1959) modified by Kates (1972). Solvents were evaporated under a stream of N_2 at 42 °C, dried, and the weight of lipid was determined gravimetrically. Total lipid extracts (10mg) were fractionated into hydrocarbons, nonpolar and polar lipid fractions on a column (1 × 20 cm) packed with silicic acid (silica gel 60F_{254}, E. Merck) with chromatography grade solvents. Hexane (25ml), Chloroform (15 ml) and methanol (15 ml) are used to isolate hydrocarbons, nonpolar lipids (except hydrocarbons) and polar lipids, respectively (Kumar and Singh, (1996). The isolated lipids in each solvent were estimated gravimetrically after evaporation of the solvents under N_2 at 42°C. The solutes were further identified by thin-layer chromatography (TLC).

Under optimum culture conditions, total nonpolar and polar lipids of *Botryococcus* species decreased by about 60 per cent from the late exponential to resting states in *B. braunii,* while hydrocarbons increased upto 70 per cent in *B. braunii* and 59 per cent in *B. protuberans* of total lipids. Radioactive carbon labelling studies also showed that contents of hydrocarbons and carotenoids increased in older cultures. In *B. braunii,* hydrocarbon ranged from 17 to 29 per cent of total nonpolar lipids with the maximum values at resting state, but triglycerols (13.5 per cent of total nonpolar lipids) appeared during the exponential state and decreased markedly at the resting state. Polar lipids were mainly made of monogalactosyl diglyceride, digalactosly diglyceride, phosphatidyl ethanolamine and phosphatidyl glycerol at each state and their highest values were noted during the exponential state. Sulfouinovosyl diglyceride, phosphatidyl inositol and phosphatidyl choline remained constant throughout the culture growth. *B. protuberans* showed similar lipid composition as those found in *B.braunii.* The contents of aldehydes and sterols (per cent of total nopolar lipid) were maximum at exponential growth state and decreased with increasing cultures age in both the species.

The decrease in polar and nopolar lipids and increase in hydrocarbons with increasing culture age in both species may be due to the conversion of major nonpolar lipid components such as fatty acids, aldehydes, triglycerols and sterols into hydrocarbons.

The reduction in algal polar lipids during the late stationary growth phase when nutrients such as nitrogen become limiting, has been reported by Piorreck and Pohl (1984). Both the species of *Botryococcus* shows higher efficiency of hydrocarbon formation during the exponential and early stationary states. Lipid accumulations in algae are chiefly esterfied fatty acids and are frequently associated with substantial increases in carotenoids while other nonsaponifiable lipids do not generally show large variations. Lipid accumulations in microalgae occur mainly during the stationary state and exhibit generally very low productivities. Fats accumulate as a result of still greater decrease in the other cell constituents (Shifrin and Chisholm, 1981). In sharp contrast, *B.braunii* and *B. protuberans* accumulate hydrocarbons mainly during the stationary and resting states. In these species the hydrocarbon productivities have been found to be higher during the exponential and early stationary states (Kumar and Singh, 1996).

It appears that *Botryococcus* behaviour differs somewhat from the one commonly reported for many microalge. Both species accumulate mostly hydrocarbons (75 per cent of the total nonpolar lipids) instead of triglycerides and this accumulation does not require blockage of cell division because it can proceed during the exponential growth state. Hydrocarbon production and accumulation are affected by the changes in physiological status of the cells. Conspicuous formation of lipids by microalgae is the result of accumulation of storage triglycerols. Many genera of Chlorophyceae, Bacillariophyceae and Xanthophyceae have oil storage as a diagnostic feature instead of or in addition to starch. With the exception of *Botryococcus* species all the algae primarily produce triglycerols, not hydrocarbons. The fundamental differences between triglycerols formation in higher plants and microalgae are that in case of algae the oil-accumulating cells are themselves fully photosynthetic. This means that the complete pathway from carbon dioxide fixation to triglycerol synthesis can be modulated by the cells.

Hydrocarbon Production Under Nitrogen Limitation and Anaerobiosis

The production and storage of lipids by microalgae are usually regulated by environmental factors and are generally species specific

(Shifrin and Chisholm, 1981). The enhancement of lipid storage during nitrogen-deficient condition has been demonstrated in other green algae (Ben-Amotz *et al;* 1985). Highest Concentration of lipid has been reported in *Monollantus salina* followed by *Nannochloropsis salina* which was deprived of nitrogen for nine days (Shifrin and Chisholm, 1981). Marked exception to this is *Dunaliella salina* (Ben-Amotz *et al;* 1985). Where there is an actual decrease in lipid content and increase in carbohydrate content under nitrogen-limiting conditions. Lipids and carbohydrates are both used as energy reserves in algae and their synthesis is regulated qualitatively and quantitatively by a number of environmental factors. Besides lipids, changes in protein, carbohydrate and fatty acids are also measured in phytoplanktons (Harrison *et al.*, 1990), and growth became limited by nitrogen deficiency. Variations in nitrogen content of the medium also cause changes in composition of carotenoids. Increasing nitrogen level in the medium leads to an increase in the biomass, protein and chlorophyll contents in freshwater green and blue-green algae (Piorreck and Pohl, 1984). In *Dunaliella* species, N-deficiency results in the accumulation of C_{40} isoprenoid beta-carotene in the chloroplast (Ben-Amotz and Avron, 1983), and in *Isochrysis* species, in the accumulation of C_{37} alkenone (Ben-Amotz *et al.*, 1985). In these algae it appears that the acyclic and isoprenod hydrocarbons are non-degradable end products that accumulate under conditions of slow growth and stress. In N-limited *Chlorella vulgaris* cells starch comprised upto 55 per cent of the dry weight (Behrens *et al.*, 1989) in comparison to 20 per cent under N-sufficient condition.

Anaerobiosis appears to trigger lipid synthesis by activating an oxygen-sensitive pyruvate dehydrogenase, which is located in the mitochondria (Inui *et al.*, 1987). Inui *et al.* (1982) showed the degradation of lipid and synthesis of carbohydrate in a green microalga, *Euglena gracilis* when cells were transferred from anaerobic to aerobic condition. Coleman *et al.* (1988) reported that under anaerobic conditions, nitrogen-deficient cells degraded carbohydrate and formed lipids by wax ester fermentation resulting in an increase in lipid content. Ben-Amotz *et al.* (1985) observed low production of hydrocarbons in green algae including *Botryococcus braunii* under optimum growth conditions and high production of hydrocarbons under growth limiting conditions. Besides nutrition, anaerobiosis also affects fatty acid and lipid composition. The accumulation of hydrocarbons and other neutral lipids is stimulated under anaerobic condition because they are incompletely oxidised products. Tornabene (1982) showed that both hydrocarbons and oxygenated compounds are synthesized through fatty acid precursors. Available literature reveals that most previous work on *Botryococcus* species has focussed upon growth, lipid and hydrocarbon production and composition under normal growth conditions (Wolf *et al.*, 1985; Vazquez-Duhalft and Greppin, 1987; Kumar and Singh, 1996) and in hydrocarbon production in relation to physiological state (Casadevall *et al.*, 1985, Singh, 2000) Little, if any, information is available concerning the lipid production of *Botryococcus* species under N-deficiency and anaerobic conditions (Singh and Kumar, 1992).

Biofertilizer

The cyanobacteria provide valuable nutrients to plants in terms of fixed nitrogen, amino acids and plant growth promoting substances which improve the soil health and texture. The role of cyanobacteria in maintaining the fertility of Indian rice field was studied by De (1939) and Singh (1961). Algalization of paddy fields is beneficial and cyanobacterial contribution to the total nitrogen in paddy fields may vary. However, it was estimated that cyanobacteria contribute 25-30 Kg nitrogen per hectare per season to the rice crop. Significant increase in yield with the application of cyanobacteria in paddy fields of Uttar Pradesh was observed (Dwivedi *et al.*, 2000). Algalization also improves phosphate solubilization (Singh *et al.*, 1981; Bisoy and Singh 1988). The application of cyanobacteria was shown to improve yield due to the production of growth promoting substances (Marsalek *et al.*, 1992). Extra cellular growth promoting substances containing amino acids have been reported in certain strains of cyanobacteria such as *Nostoc muscorum* and *Haplosiphon fontinalis* (Wikstrom *et al.*, 1997). Because of these properties strain improvement was attempted in cyanobacteria with a view to augment their nitrogen fixing potential (Tiwari *et al.*, 1991; 2001).

Food and Food Additives

The cyanobacterium *Spirulina* has traditionally been harvested from African lakes and consumed in sauces at the rate of some 9-13 grams per meal. The protein level of *Spirulina* can be as high as nuts, grains and soybeans, ranging form 50-70 per cent of algal dry weight, and shows promise for use as a protein supplement for malnourished populations. Because the cell walls are composed of mucopolysaccharides rather than cellulose, *Spirulina* is more digestible than some other microalgae. *Spirulina* is also naturally high in B vitamins and essential unsaturated fatty acids as well as high levels of β-carotene, which is converted to vitamin A during digestion. Vitamin A can prevent xerothalmia, a form of blindness that arises in malnourished children. *Spirulina* is commercially cultivated in shallow raceways, 500-5000 m^2 in area, while being stirred with paddles.

Neutraceuticals

Some strains of cyanobacteria such as *Nostoc* and *Anabaena* have already been reported to be good for human consumption. *Spirulina* and *Nostoc* have been used as a source of protein and vitamin for humans and animals (Ciferri, 1983; Kay, 1991; Gao, 1998; Takanaka *et al.*, 1998). Because of the high protein content of *Spirulina* it is used as a dietary supplement. It is also a rich source of beta-carotene, thiamine and riboflavin and is commercially available in the market. M/s. Dabur India Private Limited is marketing *Spirulina* tablets under the trade name "Sunova". In Philippines and China *Nostoc flagelliforme* and *Nostoc commune* are used as delicacy (Martinez, 1988; Gao, 1998; Takanaka *et al.*, 1998). The planktonic cyanobacterium *Spirulina platensis* is gaining increasing attention due to its nutritional and medicinal properties. Protein extracts of *S. platensis* fractions showed antioxidant activities and inhibited microsomal lipid peroxidation (Pinero *et al.*, 2001). Cyanobacteria contain significant quantities of lipids with a composition similar to vegetable oils. Many of these cyanobacterial fatty acids are essential components of human and animal diet as well as important feed additives in aquaculture (Borowitzka, 1988). The lipids of some cyanobacteria are rich in essential fatty acids (Singh *et al.*, 2002). Carotenoids are widely used as natural colorant for food, drug and cosmetic products. Shukla and Kashyap (2003) assessed the biopotential of few Antarctic cyanobacterial strains with respect to carotenoid production and suggested that these strains have a potential for commercial production of carotenoids. In India Murugappa Chettiar Research Centre, Chennai and Thapar Institute of Science and Technology, Patiala is involved in the production of *Spirulina* biomass on a large scale using open pond system of cultivation.

Biosorption of Heavy Metals

Microalgae exhibit mechanisms for the removal of free metal ions from waters, thereby both detoxifying and remediating the water. A number of microalgal strains the green algal species *Chlorella vulgaris, Scenedesmus* sp., and *Chlorococcum* sp., and the blue-green algal species, *Phormidium*, and *Spirulina*), which are potentially suitable for Hg(II), Cd(II) and Pb(II) removal in aqueous solution. Among these, unicellular alga, *Chlamydomonas reinhardtii* has only recently gained greater attention for the treatment of heavy metal contamination in aqueous solutions. Because of its wide range of heavy metal tolerance, this species can sequester an array of trace metals including: copper, zinc, lead, cadmium, cobalt, nickel, mercury, silver, and gold. In addition, it can be genetically engineered so as to enhance its ability to selectively sequester toxic heavy metals (Perales-Vela *et al.*, 2006; Tuzun *et al.*, 2005). All these organisms are reported for the hydrogen production with the ability to sequester the above said heavy metals. Another hydrocarbon rich green microalga, *Botryococcus braunii* is also reported for the biosorption of heavy metals like chromium, arsenic, cadmium and lead in considerable amounts (Metzer *et al.*, 2005).

Pharmaceuticals and Fine Chemicals

The medicinal value of cyanobacteria was appreciated as early as 1500 BC, when strains of *Nostoc* were used to treat gout, fistula and several forms of cancer (Pietra, 1990). People in Aztec civilization used to consume *Spirulina* pills four hundred years ago. In traditional as well as in modern medicine there are few appreciated algal preparations and specified bioactive compounds. However, with the availability of powerful biotechnological tools and techniques cyanobacterial biotechnology reached newer heights in terms of process and product development. Phycobiliproteins and carotenoids are effective natural blue colorants used commercially (Emodi, 1978). *Spirulina platensis* is a blue green microalga which can produce large quantities of high value products such as phycocyanin (Chen *et al.*, 1996). Biologically active compounds from *Lyngbya majuscula* showing molluscoidal activity were isolated and characterized (Orjala *et al.*, 1995; Orjala and Gerwick 1996). A study conducted by Moore (1996) revealed novel biologically active compounds from cyanobacteria which could be exploited as potentially rich sources of natural products. Novel extra cellular diterpenoids with biological activity from *Nostoc commune* have been isolated by (Jaki *et al.*, 2000). Over 40 different Nostacales species, mainly *Anabaena* and *Nostoc* sp. having activities such as anti-HIV, anticancer, antifungal, antimalarial and antimicrobial have been reported (Burja *et al.*, 2001). Among the several strains with bioactivity *Nostoc ellipsoporum* was found to show potential activity against several immunodeficiency viruses such as HIV-1, M- and T-tropic strains of HIV-1, HIV-2, SIV and FIV due to a 101 amino acid protein Cyanovirin (CV-N, cyanovirin-N). Many peptide metabolites with bioactivity have been reported. Examples are cyanopeptolin, microviridin or micropeptin produced by bloom forming freshwater cyanobacteria (Neilan *et al.*, 1999; Tillett *et al.*, 2000; Kurmayer *et al.*, 2004). Marine cyanobacterial forms are richest sources of novel bioactive compounds including toxins with pharmaceutical applications (Gustafson *et al.*, 1989; Raghavan *et al.*, 2002).

An important enzyme in the molecular biology, the restriction endonucleases have been isolated from cyanobacteria (Murray *et al.*, 1976). Isotopically labeled cyanobacterial metabolites are commercially available (Patterson, 1996). More than 200 cyanobacterial restriction endonucleases have been reported (Roberts and Macelis, 2001). Saravanan *et al.* (2003) identified and characterized a new type II restriction endonuclease from a nonheterocystous cyanobacterium *Oscillatoria foreaui*. These results have prompted scientists to exploit cost effective cultivation of cyanobacterial strains in mass culture conditions for the production of such compounds. In addition, cyanobacteria are a rich source of several polyols, polysaccharides, lipids, fatty acids, halogenated compounds and have properties to be employed as flocculants, surfactants and others (Becker, 1994).

Cyanobacteria have also been used in the control of certain bacterial diseases on plants. Pandey and Pandey (2002) reported antibacterial property of three strains of cyanobacteria such as *Lyngbya majuscula*, *Microcystis aeruginosa* and *Plectonema boryanum* and it's efficacy in controlling leaf spot disease of chilli due to *Xanthomonas vesicatoria*. This has opened new vistas for eco-friendly control of diseases using a broad spectrum of biologicals isolated from cyanobacteria against the conventional pesticides which are otherwise hazardous and unsafe to plants as well as animals.

Most of the cyanobacteria contain a wide range of products which are important industrially and pharmacologically. However, a full documentation of such potential of cyanobacteria is yet to be done. However, with the recent advancement in biotechnology and renewed interest in natural product chemistry of cyanobacteria it is expected that many more products will be identified and synthesized from them. Bioprospecting of cyanobacteria is a fast emerging discipline with potential economic implications. These organisms are undoubtedly important in the Indian context because of the economic wealth they are expected to generate in coming days in the biotechnological sphere.

Acknowledgements

The author is grateful to University Grants Commission, New Delhi for financial support and also to the Principal, Govt. P.G. College, Rishikesh (Dehradun) for encouragement.

References

Anand N (1989) Hand book of Blue Green Algae. Bishen Singh Mahendra Pal Singh, Dehradun, 1-74.

Becker EW (1994) Microalgae- Biotechnology and Microbiology, Cambridge University Press, United Kingdom.

Behrens PW, Bingham SE, Hoeksema SD, Cohoon DL and Cox JC (1989) Studies on the incorporation of CO_2 into starch by *Chlorella vulgaris.* J Appl Phycol 1: 123-130.

Ben-Amotz A and Avron M (1983) Accumulation of metabolites by halotolerant algae and its industrial potential. Ann Rev Microbiol 37: 95-119.

Ben-Amotz A, Tornabene TG and Thomas WH (1985) Chemical profile of selected species of microalgae with emphasis on lipids. J Phycol 21: 72-81.

Benemann JR, Augenstein DC and Weissman JC (1982a) "Microalgae as a source of liquid fuels, appendix: technical feasibility analysis." *Final Report,* U.S. Department of Energy, unpublished, pp 126.

Benemann JR, Goebel RP, Weissman JC and Augenstein DC (1982b) "Microalgae as a source of liquid fuels." *Final Report,* U.S. Department of Energy, pp 202.

Benemann JR, Weissman JC, Goebel RP and Augenstein DC (1986) Microalgae fuel economics. In: Algal Biomass Technologies. Barclay WR and McIntosh RP (eds), Nova Hedwigia Beih, 83, *J. Cramer, Berlin*, pp. 176-191.

Bisoy RN and Singh PK (1988) Effect of phosphorous fertilization in blue-green algal inoculum production and nitrogen yield under field conditions. Biol Fert Soils 5: 338-343.

Bligh EG and Dyer WJ (1959) A rapid method of total lipid extraction and purification. Can J Biochem Physiol 37:911-917.

Borowitzka MA (1988) Fats, oils and hydrocarbons. *In*: Borowitzka MA Borowitzka, LJ (eds), Microalgal Biotechnology. Cambridge University, Press, Cambridge, 257-287.

Broady PA, Garrick R and Anderson GM (1996) Diversity, distribution and dispersal of Antarctic terrestrial algae. Biodiversity Conservation 5: 1307-1335.

Burja AM, Banaigs B, Abou-Mansour E and Burgess JG (2001) Marine cyanobacteria- a prolific source of natural products. Tetrahedron 57: 9347-9377.

Casadevall E, Dif D, Largeau C, Gudin D, Chaumont D and Desanti O (1985) Studies on batch and continuous cultures of Botryococcus braunii: A hydrocarbon production in relation to physiological state, cell ultrastructure and phosphate nutrition.Biotechnol Bioeng 27:286-295.

Chen F, Zhang Y and Guo S. (1996) Growth and phycocyanin formation of *Spirulina platensis* in photoheterotrophic culture. Biotechnol Lett 18: 603-608.

Ciferri O (1983) *Spirulina*, the edible microorganism. Microbiol Rev. 47: 551-578.

Cohen Z (1986) Products from microalgae. In: Hand book of Microalgal Mass culture. Richmond A (ed), CRC Press, Boca Raton, FL,421-454.

Coleman LW, Rosen BH and Schwartzbach SD (1988) Environmentral control of carbohydrate and lipid synthesis in *Euglena*. Plant cell Physiol 29:423-432.

De PK (1939) The role of blue green algae in nitrogen fixation in rice fields. Proc Roy Soc 127: 121-139.

Desikachary TV (1959). Cyanophyta. Indian Council of Agricultural Research, New Delhi.

Dwivedi CP, Sinha SK, Ashraf SS, Verma DC and Dubey AK (2000) Application of blue green algae (BGA) fertilizer in paddy cultivation under farm conditions in Uttar Pradesh.

Emodi A (1978) Carotenoids: Properties and Applications. Food Technol 32: 38-42.

Gao K (1998) Chinese studies on the edible blue-green algae *Nostoc flagelliforme*: a review. J Appl Phycol 10: 37-39.

Gaur JP and Rai LC (2001) *In*: Algal adaptation to environmental stresses, Rai LC and Gaur JP (eds), Springer, Berlin, 363-388.

Grant WD and Tindall WJ (1986) *In*: Microbes in extreme environments, Herbert TA and Codd GA (eds), Academic Press, London, 25-54.

Gustafson KR, Cardellina JH, Fuller RW, Weislon OS, Kiser RF and Snader KM (1989) Antiviral sulfolipids from cyanobacteria (blue-green algae). J Nat Cancer Inst 81: 1254.

Hall DO (1998) Immobilized photosynthetic system for the production of fuels and chemicals. In: 8Th International Biotechnology Symposium, Paris, July 17-22, Abstrak Book 34.

Hallen back PC, Kochain LV, Weissman JC and Benemann JR (1978) Solar energy conversion with hydrogen producing cultures of the blue green algae Anabaena cylindrical. *Biotechnol. Bioeng.* Symp. 8: 283-297.

Harrison PJ, Thompson PA and Caderwood GS (1990) Effects of nutrients and light limitation on the biochemical composition of phytoplankton. J Appl Phycol 2: 45-56.

Haselkorn R (1978) Heterocysts. Ann Rev Plant Physiol 29: 319-344.

Inui H, Ono K, Miyatake K, Nakano Y and Kitaoka S (1987) Purification and characterization of pyruvate:NADP$^+$ oxidoreductase in *Euglena gracilis*. J Biol Chem 262:9130-9135.

Inui H, Miyatake K, Nakano Y and Kitaoka S (1982) Wax ester fermentation in *Euglena gracilis*. FEBS letter 150:89-93.

Jaki B, Orjala J, Heilmann H, Linden A, Vogler B and Sticher O (2000) Extracellular diterpenoids with biological activity from the cyanobacterium *Nostoc commune*. J Nat Prod 63: 339-343.

Karube I, Ikemoto H, Kajiwawa K, Tamiya E and Matsuoka H (1986) Phytochemical energy conversion using immobilized blue-green algae. Journal of Biotechnology 4: 73.

Kates M (1972) Isolation, analysis and identification of lipids *In*: Work TS and Work E (eds),Techniques in Lipidology. Elsevier, Amsterdam, 268-618.

Kay RA (1991) Microalgae as food and supplement. Crit Rev Food Sci Nutr 30: 555-573.

Kumar HD and Singh Y (1996) Lipid profile and hydrocarbon production by Botyococcus. Current Res Plant Sc 2:71-80.

Kumar, H.D. and Singh 1995 A Text Book of Algae. Affiliated East West Press. New Delhi.

Kumazawa S and Mitsui A (1982) Hydrogen metabolism of photosynthetic bacteria and algae. In: CRC Handbook of Biosolar Resources. (ed) Zaborsky, Vol. 1. Part 1, CRC Press Inc., Boca Raton, Florida.

Kurmayer R, Christiansen G, Fastner J and Borner T (2004) Abundance of active and inactive microcystin genes in populations of the toxic cyanobacterium *Plankthrix* sp. Environ Microbiol 6: 831-841.

Lind blad P, Oxelfelt F, Tamagini P and Troshino O (1998) Cyanobacterial Biotechnology Nostoc PCC 73102 and H_2: knowledge, research and biotechnological challenges. In: Cyanobacterial Biotechnology. (eds) Subramanian, G., Kaushik, B. D. and Venkataraman, G.S., *Oxford and IBH Publishing Co. Pvt. Ltd.*, Borowitzka (eds.), Micro-Algal Biotechnology, Cambridge University Press, New Delhi, 1-5.

Marsalek B, Zahradnickova H and Hronkova M (1992). Extracellular abscisic acid production by cyanobacteria under salt stress. J Plant Physiol 139: 506-508.

Martinez MR. (1988) *Nostoc commune* Vauch. A nitrogen fixing blue-green algae, as source of food in the Philippines. Philippine Naturalist. 71: 295-307.

Metzger P and Largeau C (2005) *Botryococcus braunii*: a rich source for hydrocarbons and related ether lipids. *Appl Microbiol Biotechnol* 66: 486-496.

Mitsui A (1980) Saltwater based biological solar energy conversion for fuel, chemicals, fertilizer, food and feed. In: *Proc. Bioenergy'*80. Bioenergy Council, Washington, D.C., 486-491.

Moore RE (1996) Cyclic peptides and depsipeptides from cyanobacteria: a review. Ind J Microbiol 16: 134-143.

Murray K, Gughes SG, Brown JS and Bruce SA (1976). Isolation and characterization of two sequence specific endonucleases from *Anabaena variabilis*. Biochem J 138: 123-127.

Neenan B, Feinberg D, Hill A, McIntosh R and Terry K (1986) Fuels from microalgae: Technology status, potential and research requirements, *Solar Energy Research institute, Golden Co.*, SERI/ SP, pp: 31-255.

Neilan BA, Dittman E, Rouhianinen L, Bass RA, Sivonen K and Borner T (1999) Non ribosomal peptide synthesis and the toxigenecity of cyanobacterium. J Bacteriol 181: 4089-4097.

Orjala J and Gerwick WH (1996) Barbamide, a chlorinated metabolite with molluscoidal activity from the carribean cyanobacterium *Lyngbya majuscula*. J Nat Prod 59: 427-430.

Orjala J, Nagle D and Gerwick WH (1995) Antillatoxin, an exceptionally ichthyotoxic cyclic lipopeptide from the marine cyanobacterium *Lyngbya majuscula*. J Am Chem Soc 117: 8281-8282.

Pandey U and Pandey J (2002) Antibacterial properties of cyanobacteria: A cost effective and eco-friendly approach to control bacteria leaf spot disease of chilli. Curr Sci 82 (3): 262-264.

Patterson GML (1996) Biotechnological applications of cyanobacteria. J Sci Ind Res 55: 669-684.

Perales-Vela HV, Pena-Castro JM and Canizares-Villanueva RO (2006) Heavy metal detoxification in eukaryotic microalgae. Chemosphere (in press)

Pietra F (1990) A Secret World: Natural Products of Marine Life. Ist ed. Birkhauser, Basel, Switzerland.

Pinero Estrada JE, Bermejo Bescos P and Villar del Fresner AM (2001) Antioxidant activity of different fractions of *Spirulina platensis* protean extract. Farmaco 56(5-7): 497-500.

Piorreck M and Pohl P (1984) Formation of biomass, total protein, chlorophylls lipids and fatty acids in green and blue-green algae during one growth phase. Phytochemistry 23:217-223.

Prescott, GW (1969) The Algae: a Review. London: Nelson.

Raghavan C, Kadlamani B, Thirunalasundari T, Subramahnian G and A Skbarsha M A (2002) A study of the male antifertility properties of the marine cyanobacterium *Oscillatoria willei* BDU 135011: A preliminary report. *In*: Proceedings of the XX symposium. Bharathidasan University, Thiruchirapalli, 137-138.

Rai AN (1990) Cyanobacteria in symbiosis. *In*: AN Rai (ed), Hand book of Symbiotic cyanobacteria CRC Press, Bocca Raton.

Rai LC, Singh S and Pradhan S (1998) Biotechnological potential of naturally occurring and laboratory grown *Microcystis* in biosorption of Ni^{2+} and Cd^{2+}. Curr Sci 74 (5): 461-464.

Roberts RJ and Macelis D (2001) REBASE-restriction enzymes and methylases. Nucleic Acids Research. 29: 268-269.

Robinson PK, Mak AL and Trevan MD (1986) Immobilized Algae: A Review. Process Biochemistry 21:122-127.

Saravanan M, Elango K, Chandrashekaran S, Anand N and Nagaraja V (2003) A new type II restriction endonuclease, *O. foreaui* from nonheterocystous cyanobacterium *Oscillatoria foreaui*. Curr Sci 85(2): 188-191.

Shifrin NS and Chisholm SW (1981) Phytoplankton lipids: Environmental influence on production and possible commercial applications. *In*: Shelef G and Soeder CJ (eds), Algae Biomass. Elsevier/North Holland Biochemical Press, Amsterdam, 627-645.

Shukla SP and Kashyap AK (2003) An assessment of biopotential of three cyanobacterial isolates from Antarctic for carotenoid production. Ind J Biochem Biophys 40: 362-366.

Singh PK, Panigrahi BC and Satpathy KB (1981). Comparative efficiency of *Azolla*, blue-green algae and other organic manures in relation to N and P availability in flooded rice soil. Plant and Soil 62: 35-44.

Singh RN (1961) Role of blue green algae in nitrogen economy of Indian Agriculture. Indian Council of Agricultural Research, New Delhi, 175.

Singh SC, Sinha RP and Hader DP (2002) Role of lipids and fatty acids in stress tolerance in cyanobacteria. Acta Protozool 41: 297-308.

Singh Y (2000) Studies on a hydrocarbon rich microalga, *Botryococcus* at different physiological states. J Indian Bot Soc 79:115-118.

Singh Y And Kumar HD (1992) Lipid and hydrocarbon production by *Botryococcus* species under nitrogen limitation and anaerobiosis. World J Microbiol Biotechnol 8:121-124.

Stal LJ (2000) Cyanobacterial mats and stromatolites. *In*: Whitton BA and Potts M (eds), The ecology of cyanobacteria. Kluwer Academic Publishers, Netherlands, 61-120.

Takanaka H, Yamaguchi Y, Sakaki S, Watarai K, Tanaka N, Hori M, Seki H, Tsuchida M, Yamada A, Nishimori T and Mrinaga T (1998). Safety evaluation of *Nostoc flagelliforme* (Nostocales, Cyanophyceae) as a potential food. Food Chem Toxicol 36: 1073-1077.

Thajuddin N and Subramanian N (2005). Cyanobacterial biodiversity and potential applications in biotechnology. Curr Sci. 89(1): 47-57.

Thajuddin N, Nagasthya A, Chelladevi R and Saravanan P (2002) Biodiversity of cyanobacteria in different saltpans of Pudukottai District, Tamilnadu. Sea weed Res Utiln 24: 1-11.

Tillet D, Dittman E, Erhard M, Von Dohren H, Borner T and Neilan BA (2000) Structural organization of microcystin biosynthesis in *Microcystis aeruginosa* PCC 7806: An integrated peptide-polypeptide synthetase system. Chem Biol 7: 753-764.

Tiwari DN, Kumar A and Mishra AK (1991) Use of cyanobacterial diazotrophic technology in rice agriculture. Appl Biochem Biotech 28: 387-396.

Tiwari DN, Mishra AK and Singh AP (2001) Transposan mutagenesis and other molecular tools for strain improvement in cyanobacteria. *In*: Recent advances in the exploitation of blue green algae, PK Singh *et al.* (eds), Venus Printers and Publishers, New Delhi, 85-94.

Tornabene TG (1982) Microorganisms as hydrocarbon producers. Experientia 38:43-46.

Tuzun I, Bayramoglu G, Yalcyn E, Basaran G, Celik G and Aryca MY (2005) Equilibrium and kinetic studies on biosorption of Hg(II), Cd(II) and Pb(II) ions onto microalgae Chlamydomonas reinhardtii. Journal of Environmental Management 77:85–92.

Vazquez-Duhalt R and Greppin H (1987) Growth and production of cell constituents in batch cultures of *Botryococcus sudeticus*. Phytochemistry 26:885-889.

Wikstrom P, Szwajeer E, Brodelius P, Nilsson KN and Mosbach K (1997) Formation of alphaketo acids from amino acids using immobilized bacteria and algae. Biotechnol Lett 4: 153.

Wolf FR Nemethy EK Blanding JH and Bassham JA (1985) Studies on the biosynthesis of unusual acyclic isoprenoids in the alga *Botryococcus braunii*. Phytochemistry 24:733-737.

Wolk CP (1996) Heterocyst formation. Ann Rev Gen 30: 59-78.

Chapter 2

In vitro Propagation of *Dalbergia* Species Towards Large-Scale Supply of Elite Planting Material for Reforestation

Pradeep K. Chand* [1] **and Kunja B. Satapathy** [2]

[1]*Post-Graduate Department of Botany, Utkal University, Bhubaneswar – 751 004, Orissa, India*

[2]*Department of Botany, Vyasanagar College, Jajpur Road – 755019, Orissa, India*

ABSTRACT

Tree legume species of *Dalbergia viz. D. sissoo* Roxb., *D. latifolia* Roxb. and *D. lanceolaria* L. are much valued for quality timber and also due to their medicinal properties. Mass propagation of these species has been a requirement towards generation of elite planting material to be used for reforestation programs. Propagation via conventional means fail to cater to this objective owing to their long breeding cycle coupled with high heterozygosity and associated problems, common to woody plant species. Micropropagation through *in vitro* culture of plant tissues has emerged as an effective means for rapid

* Corresponding Author: Tel: (O) 91-674–2581598, (R) 91-674–2583529, (M) 9861356080; Fax: 91-674-2545993; E-mail : pkchanduubot@rediffmail.com, pkchanduubot@yahoo.co.in

clonal multiplication of selected woody trees despite inherent recalcitrance. In view of this, a review is presented herein to document the progress made hitherto at the Post-Graduate Department of Botany, Utkal University towards resulting in reproducible protocols for *in vitro* propagation of all the three economically important *Dalbergia* species as aforesaid.

Keywords: *Tree legumes, Dalbergia, Micropropagation, Reforestation.*

Abbreviations

BA: ^{6}N-benzyladenine; 2,4-D: 2,4-dichlorophenoxyacetic acid; GA_3: gibberellic acid; IAA: indole-3-acetic acid; IBA: indole-3-butyric acid; IPA: indole-3-propionic acid; 2-iP: 2-isopentenyladenine; Kin: kinetin; NAA: α-naphthalene acetic acid; TDZ: thidiazuron; ZR: zeatin riboside.

Introduction

India is known as the world's botanical garden because of its rich biodiversity being one of the world's top twelve mega-diversity countries (Singh and Choudhery, 2002). Besides, encompassing ten bio-geographic regions (Rodgers and Panwar, 1990), it has over 40 sites which are known for their high order of vegetational endemism and genetic diversity because of varied climatic and altitudinal habitats (Nayar, 1996). Unfortunately, forest is losing its natural flora due to shrinkage of the forest lands at an alarming rate *i.e.* about 1.5 million hectares every year and leaving only 8 per cent forest cover, at present, as against a mandatory 33 per cent of the geographical area (Sivarajan and Balachandran, 1999). This phenomenal loss largely owes to excessive human interference and natural calamities. Human activities for development include construction of new roads through forest areas, acquisition of forest lands for agriculture and industries and rapid urbanization through rehabilitation programmes. The biodiversity is threatened further due to frequent landslides, super-cyclones or other natural calamities. If such activities are continued it is apprehended that we may lose a large number of species of useful forest plants within a coming few years. This would cause a tremendous adversity to the human survival and sustenance. To mitigate the situation various measures are contemplated which include commercial cultivation, habitat conservation, setting up of natural reserves such as botanical gardens, sanctuaries, national parks and biosphere reserves coupled with implementation of laws for restricting export of plants (Rao *et al.*, 2003).

Forest trees are not only valued for wood but also for a range of other natural products such as fibres, alkaloids, tannins and resins. In order to meet the high market demand for wood and other natural forest products, many of the trees have been indiscriminately logged. This unregulated felling of trees from the natural habitat and inadequate efforts for their replenishment has resulted in a marked decline in their population. Conventional methods of propagating trees either through seeds, root suckers, cuttings or graftings are beset with many problems. None of these methods is very efficient in producing sufficient numbers of planting material. While production of root suckers in trees is age-dependent, propagation through cuttings is not only time consuming and tedious but majority of cuttings are also difficult-to-root. For several plants the percentage of seed viability is very poor and germination is very low. In addition, seed-derived plants will not exhibit absolute trueness-to-type vis-à-vis the donor with regards to desirable properties. Tree improvement by conventional breeding is a slow process because of their long juvenile period coupled with high heterozygosity.

Micropropagation is an aspect of plant tissue culture which has been most convincingly documented with regard to its feasibility for commercial application. Consequently it has been extensively used for rapid and large scale propagation of a number of plant species. The economical and ecological importance of leguminous forest trees necessitates the application of this technique for their clonal multiplication (Tomar and Gupta, 1988). Unfortunately, regeneration of forest trees, in general and legumes in particular, through tissue culture has been a difficult problem. Nevertheless, during the past few years tissue culture techniques have been extensively used for large-scale propagation of a number of forest trees (Bajaj, 1986). Besides, selected woody legumes have been successfully propagated *in vitro* using juvenile as well as matured plant parts (Trigiano *et al.*, 1992). However, it has not been possible as yet to regenerate several woody plant species including some economically important leguminous forest trees due to a genetically driven *in vitro* recalcitrance (McCown, 2000).

The genus *Dalbergia* of family Fabaceae comprises of several important timber-yielding forest trees legume species of India of which *D. sissoo* Roxb., *D. latifolia* Roxb. and *D. lanceolaria* L. are noteworthy. Because of its great strength, durability and elasticity, the wood of *D. sissoo* (Sissoo) is used for making furniture, cabinets, musical instruments, ornamental veneers and high quality commercial plywood. Carefully selected and manufactured Sissoo plywood also meets the specifications of aircrafts. The leaves are used as fodder and green manure. In addition, a light-brown highly viscous oil is obtained from the heartwood and is used as a lubricant for heavy machinery (Anonymous, 1970). Wood is useful in leprosy, boils, eruptions and to allay vomiting. Roots are astringent and leaf decoction is useful in gonorrhoea (Chopra *et al*, 1956). *Dalbergia latifolia* Roxb. (East Indian Rosewood) is valued for its wood which is used for carving, pattern making, calico-printing, mathematical instruments and ornamental poly-boards (Anonymous, 1970). Root, bark and leaf are useful in polyurea, sciatica, chronic ulcer, leprosy, leucoderma, urinary bladder disorders, burning sensations, and oedema (Yoganarasimhan, 2000). *Dalbergia lanceolaria* L. yields timber that is used for tool handles and agricultural implements. Its wood is also suitable for carving, boarding, rafters, scantlings, packing cases and general construction purposes. The plant is also important for its medicinal value. The decoction of the bark, that contains 14 per cent tannin, is used in dyspepsia while the seed oil is recommended for rheumatic affections (Anonymous, 1970).

In view of the proven economic use of the three important *Dalbergia* species as aforesaid it was imperative to develop *in vitro* culture protocols for rapid and large scale propagation for each one of these important plant species aiming at generation of sufficient planting material so as to facilitate a successful reforestation program. *In vitro* propagation of *D. sissoo* Roxb. using nodal segments (Datta and Datta, 1983; Datta *et al.*, 1983; Gulati and Jaiwal, 1996) or axillary buds (Dwara *et al.*, 1984) derived from mature trees, organogenesis from callus cultures (Sharma and Chandra, 1988) or somatic embryogenesis from semi-mature zygotic embryos (Das *et al.*, 1997) has been reported. Most of these reports, however, were preliminary in nature, and the data provided were inadequate. Information on frequency of shoot proliferation, shoot number, shoot length, percentage of rooting and number of roots per shoot was missing. This underpinned a systematic investigation aimed at developing reproducible protocols for whole plant regeneration of all the three important *Dalbergia* species, as aforesaid, by attempting to exploit different *in vitro* culture methods.

Materials and Methods

Seed Germination and Establishment of Shoot Cultures from Cotyledonary Nodes

A mature elite tree for each one of the *Dalbergia* species *viz. D. sissoo* Roxb., *D. latifolia* Roxb. and *D. lanceolaria* L. was selected from which the dried and mature pods were collected. Healthy seeds were removed from the pods, washed with running tap water (15 min), immersed in an aqueous solution of 5 per cent liquid detergent, Laboline (Qualigens, India) for 5 min and rinsed with distilled water (5-6 changes). The seeds were surface-disinfected with an aqueous solution of 0.1 per cent $HgCl_2$ for 8 min followed by five or six rinses in autoclaved distilled water. Thereafter, seeds were transferred to screw-capped glass tubes (60 ml, Borosil, India) containing growth regulator-free half-strength culture medium with 15 g l^{-1} sucrose, 50 mg l^{-1} myo-inositol and 8.0 g l^{-1} agar (Bacteriological grade, Hi-media, India). Four different basal media formulations *viz.* MS (Murashige and Skoog, 1962), BM (Blaydes, 1966), B_5 (Gamborg *et al.*, 1968) and WPM (Lloyd and McCown, 1981) were tested. The pH of all media was adjusted to 5.8 prior to autoclaving at 120 °C and 104 kPa for 20 min. The culture tubes were incubated at 25±1 °C, under light at a 35 µmol m^{-2} s^{-1} photon flux density (PFD) provided by white-fluorescent tubes and 60 per cent relative humidity. Seven-day-old axenic seedlings served as the source of explants. After removal of the radicle and primary shoot from the seedlings, cotyledonary nodes were inserted vertically into 200 ml screw-capped glass jars containing full-strength culture medium with 30 g l^{-1} sucrose, 100 mg l^{-1} myo-inositol, and augmented with either of the cytokinins (BA, Kin, 2-iP, ZR) each at 0.5-5.0 mg l^{-1} or a cytokinin-like substance, TDZ at 0.01-0.1 mg l^{-1} and being gelled with 8 g l^{-1} agar. The cultures were maintained under conditions similar to those described earlier for seed germination.

The original cotyledonary nodes were repeatedly subcultured on fresh shoot multiplication medium (MS+2.0 mg l^{-1}BA) after each harvesting of the newly formed shoots. The primary shoots obtained from each harvest at an interval of 30 days were cut into single node pieces (2-3 nodes/shoot) and transferred to MS medium containing 1.0 mg l^{-1} BA for axillary shoot proliferation.

Culture of Semi-mature or Mature Cotyledons for Adventitious Regeneration

For semi-mature cotyledons (60-75 days following pollination), healthy green pods were collected from a mature elite tree of each of the three *Dalbergia* species, washed under running tap water (30 min), treated with an 0.1 per cent (v/v) aqueous solution of Tween 20 (Hi-media, Mumbai, active ingredient-Polysorbitol 20) for 15 min and washed once with distilled water followed by 0.06 per cent (v/v) Savlon antiseptic solution (Johnson and Johnson, USA, active ingredients-chlorhexidine gluconate and cetrimide) for 15 min followed by 5-6 rinses with distilled water. Pods were then immersed in 70 per cent ethanol (1 min) and surface-disinfected with an aqueous solution of 0.1 per cent $HgCl_2$ (25 min), and washed with autoclaved distilled water (3-4 changes). Pods were cut aseptically and the cotyledons excised after removing the embryonic axis. Cotyledons (2-3 explants/vessel) were transferred to 300 ml screw-capped glass jars containing 25 ml culture medium supplemented with 30 g l^{-1} sucrose, 100 mg l^{-1} myo-inositol, 8.0 g l^{-1} agar (Bacteriological grade, Hi-media, Mumbai), and augmented with 0.5, 1.0, 2.0 and 3.0 mg l^{-1} BA alone or in combination with 0.05 mg l^{-1} NAA. Cultures were maintained at 25±2 °C under a 16/8 h light/dark photoperiod, and 40 µmol m^{-2} s^{-1} PFD provided by cool white fluorescent tubes. Newly formed shoots (1.0-1.5 cm) along with the original explant were transferred to MS medium (concentration of macro-elements reduced to half) containing 0.5, 1.0, 2.0 and 3.0 mg l^{-1} BA alone or with 0.05 mg l^{-1} NAA for shoot elongation. Cultures were transferred to a fresh medium after intervals of 4 weeks.

Mature cotyledons excised from 8-10 day-old axenic seedlings were transferred to 300 ml screw-capped glass jars containing full-strength culture medium supplemented with 1.0, 2.0, 3.0 and 4.0 mg l^{-1} BA (control treatments). A set of cotyledons was given a pre-treatment for 24, 48 or 72 h with 2.0 mg l^{-1} BA in liquid MS medium. Following pre-treatment, the cotyledons were transferred to regeneration medium (MS medium supplemented with 1.0, 2.0, 3.0 and 4.0 mg l^{-1} BA and 8.0 g l^{-1} agar). Cultures were maintained under similar conditions as described earlier for semi-mature cotyledons.

Callus Initiation and Proliferation

Primary roots, hypocotyls, cotyledons, stem and leaflets from 10-12 day-old seedlings were inoculated on different basal media formulations (MS, BM, B_5 and WPM) supplemented with any of the auxins (IAA, NAA, 2,4-D) and a cytokinin (BA) either singly or in various combinations. Explants (tender internodes and leaflets) from young twigs of the same mature tree, from which the seeds were collected, were surface disinfected as for seeds and inoculated on the above medium. Auxin concentrations ranging from 1.0-4.0 mg l^{-1} and BA from 0.25-1.0 mg l^{-1} were used for callus induction and proliferation. The pH of all media was adjusted to 5.8 which were solidified with 8.0 g l^{-1} agar. The cultures were maintained at 25±1°C under 16 h photoperiod with a PFD of 35 µmol $m^{-2}s^{-1}$ as provided by cool white fluorescent lamps. Callus growth was measured by subculturing pre-weighed calli followed by harvesting at one week intervals for fresh weight and dry weight determination.

Shoot Organogenesis in Callus Cultures

After 2-6 weeks of culture small portions (*ca.* 0.5 cm diam) from fast growing calli were transferred to full-strength culture medium augmented with BA (1.0-3.0 mg l^{-1}) alone or in combination with either NAA or IAA (0.1-0.5 mg l^{-1}) for eliciting shoot regeneration from callus cultures. The PFD was increased to 50 µmol $m^{-2}s^{-1}$ for facilitating shoot organogenesis.

Initiation of Cell Suspension Cultures

Friable hypocotyl-derived callus from the second subculture passage was transferred to 250 ml Erlenmeyer flasks each containing 50 ml liquid culture medium with composition as above for callus culture but without agar. Cultures were placed on a horizontal rotary shaker and agitated (80-100 rpm) under the same environmental conditions as described above. Cells were harvested after 10 d by sieving the cell suspension through a 500 µm pore size sieve. Smaller cell clumps, collected below in a beaker, were transferred to fresh medium of the same composition for further proliferation at initial inoculum densities of 0.5, 1.0 and 2.0 per cent (settled cell volume, SCV). A non-invasive method was used to measure the growth of the cell-suspension culture. A culture flask fitted with a graduated side arm was tilted to collect 10 ml suspension in the side arm

(Figure 2.9a), the cells were allowed to settle, and height of the sediment (SCV) was recorded. The change in the volume of the cell aggregates that settled represented the growth of the cell suspension. This was repeated at 2 d intervals. Suspension cultures were maintained by subculturing at 10-12 d intervals. The viability of the cell suspension was checked every 2 wk using fluorescein diacetate (FDA) (Larkin, 1976). Viable cells exhibited yellow-green fluorescence (495-565 nm) due to UV excitation of fluorescein.

Microcolonies (15-20 cells) were harvested from the cell-suspension culture by sieving and thereafter embedded in MS medium containing 2.0 mg l^{-1} NAA and 0.5 mg l^{-1} BA. Three different gelling agents were used for embedding the microcolonies: agar (5.0 g l^{-1}; bacteriological grade, Hi-media); sodium alginate (10 g l^{-1}; Sigma, St. Louis, MO, USA) and Phytagel (1.2 g l^{-1}; Sigma). Media gelled with agar or Phytagel were molten to embed the microcolonies. Plating efficiency was determined after 25 d based on the percentage of microcolonies that formed larger colonies (macrocolonies, 2-3 mm diam).

Shoot Organogenesis from Cell Suspension-Derived Callus

Larger colonies (2-3 mm) were transferred to fresh medium of the same composition for further proliferation. Finally the callus was transferred to 300 ml screw-capped glass jars containing culture medium supplemented with 1.0-5.0 mg l^{-1} BA and 0.25-0.5 mg l^{-1} NAA for shoot bud differentiation.

Rooting of Shoots and Transfer of Plantlets to Soil

Two basal media formulations *i.e.* MS and WM (White, 1963) supplemented with auxins (IAA, IBA, IPA, NAA), were used for inducing root development from excised regenerated shoots. Shoots, 3-4 cm in height with two or three leaflets, derived from secondary cultures were transferred to half-strength MS or White's medium each containing 50 mg l^{-1} myo-inositol, 20 g l^{-1} sucrose and 7.0 g l^{-1} agar. The medium was augmented with 0.5-2.0 mg l^{-1} IAA, 0.5-2.0 mg l^{-1} IBA and 0.5-2.0 mg l^{-1} IPA either each individually or in combinations. The rooted shoots were transferred to half-strength MS for further elongation of the roots.

Well-rooted shoots were removed from the culture medium, and the roots were washed gently with water to remove agar. Plantlets were then transferred to small plastic pots (7.5 cm diam) containing autoclaved vermi-compost (Ranjan's Agrotech, Bhubaneswar) or Soil rite mix (Karnataka Explosives, Bangalore, India) and kept in a plant growth chamber at 26 °C with illumination at 50 µmol $m^{-2} s^{-1}$ photosynthetic photon flux density (PPFD) provided by incandescent bulbs. The relative humidity was set originally at 85 per cent which was reduced gradually and the plantlets were removed from the growth chamber after 3 wk. Plantlets were then transferred to larger earthenware pots (18 cm diam) containing soil: compost (1:1) and kept in the shade for another 2 wk before transfer to outdoors under full sun.

Results

Multiple Shoot Proliferation in Cotyledonary Nodes from *in vitro*–Germinated Seeds

About 98-100 per cent of seeds of all the three *Dalbergia* species studied, *i.e. D. lanceolaria, D. latifolia* and *D. sissoo*, germinated within 3-4 days of inoculation on a growth regulator-free half-strength MS medium. Cotyledonary nodes excised from 7-day-old axenic seedlings were used as explants, but none of these, regardless of species, responded morphogenically to the culture medium denied of cytokinin. Multiple shoots were induced only when the medium was supplemented with either of a cytokinin such as BA, Kin, 2iP, ZR or a cytokinin-like substance, TDZ. The frequency of axillary shoot proliferation and the number of shoots per explant increased with increasing concentration of cytokinins up to an optimum level. Of all the basal formulations tested (MS, BM, B_5, and WPM), MS was most responsive and of all the cytokinins examined, BA was most effective. BA at an optimal level of 2.0 mg l^{-1} induced multiple shoot buds in 99 per cent of cotyledonary node explants within 4-6 days in *D. sissoo*, 93 per cent within 5-7 days in *D. latifolia* and 92 per cent within 6-8 days in *D. lanceolaria*. The buds appeared as small green protuberances that elongated into slender shoots possessing diminutive leaves (Figures 2.1a and b) (Pradhan *et al.*, 1998 a,b; Pradhan, 2000). In 30 days from each explant, 10-12 shoots in *D. latifolia*, 7-8 shoots in *D. sissoo* and 5-6 shoots in *D. lanceolaria* were developed, shoots averaging 4.2-4.5 cm in length. Axillary shoot

Figure 2.1: Multiple Shoot Proliferation in Cotyledonary Node Explants Cultured in MS + 2.0 mg l^{-1} BA Followed by Rooted Shoots (Plantlets) in Culture Tubes Containing MS + 1.0 mg l^{-1} IAA + 1.0 mg l^{-1} IBA + 1.0 mg l^{-1} IPA. a and b: *Dalbergia latifolia*; c and d: *Dalbergia sissoo*

buds were induced in the cotyledonary nodes within 10-12 days on MS containing 2iP, Kin or ZR; however, the frequency of shoot proliferation was relatively low and shoots per explant were fewer as compared to that recorded on BA containing media. The frequency of axillary shoot proliferation declined when the concentration of BA, Kin, 2iP and ZR was enhanced beyond 2.0 mg l^{-1} and shoots induced at higher cytokinin concentrations (3.0-5.0 mg l^{-1}) failed to elongate. Of all the cytokinin types tested TDZ was least effective in inducing shoot proliferation. Although multiple shoots were induced on TDZ augmented medium the shoots failed to elongate and were often fasciated.

Shoot cultures were multiplied by repeatedly subculturing the original cotyledonary nodes on the shoot multiplication medium MS+BA (2.0 mg l^{-1}) after each harvesting of the newly formed shoots. The original cotyledonary nodes could be subcultured twice without affecting the shoot forming potential. In 30 days from each explants 6-8 shoots of *D. sissoo*, 8-10 shoots of *D. lanceolaria* and 10-12 shoots of *D. latifolia* were obtained. The primary shoots obtained after each harvest were multiplied as stem nodal segments on MS containing 1.0 mg l^{-1} BA and 80 per cent of these nodes produced multiplied shoots (2 or 3 shoots per node) in 25-30 days. Although multiple shoot buds were induced in the nodal explants on medium containing higher level of BA (2.0 mg l^{-1}), shoot elongation was inhibited. The nodal explants could be multiplied further for 3 passages after which there was a decline in the shoot proliferation. In 90 days from a single cotyledonary node about 60-70 shoots were obtained in *D. sissoo*, 100 shoots in *D. lanceolaria* and about 120 shoots in *D. latifolia*. (Pradhan *et al.*, 1998 a, b; Pradhan, 2000).

Adventitious Shoot Organogenesis in Semi-mature and Mature Cotyledons from *in vitro*–Germinated Seeds

Semi-mature Cotyledons

Shoot buds were differentiated from the proximal end of the semi-mature cotyledons (towards embryo attachment site) of all the three *Dalbergia* species in MS medium containing 0.5-3.0 mg l^{-1} BA alone or with 0.05 mg l^{-1} NAA. MS medium containing 1.0 mg l^{-1} BA and 0.05 mg l^{-1} NAA induced adventitious shoots in 52 per cent of the cotyledons within 15-20 days and from each cotyledon an average of 4.1 shoot buds developed. These values were significantly different from those obtained with remaining treatments (Figure 2.2; Singh *et al.*, 2002). At higher concentrations of BA, both the frequency of shoot bud formation and mean number of shoots per explants significantly decreased. Proliferation of shoots occurred on MS

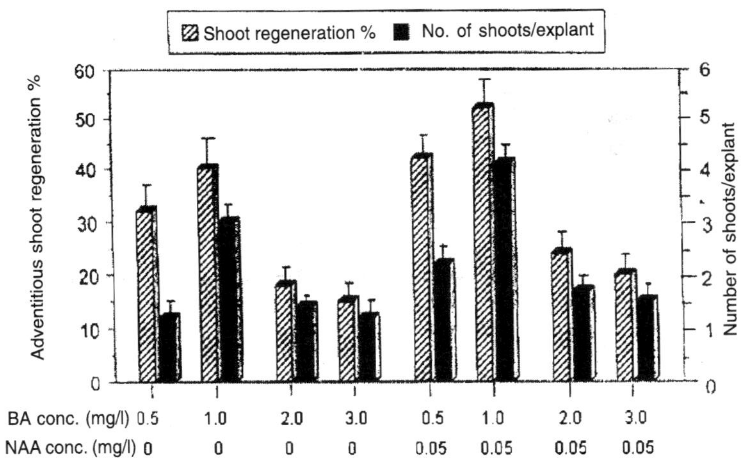

Figure 2.2: Percentage Response for Adventitious Shoot Regeneration and Mean Number of Shoots per Explant from Semi-Mature Cotyledons of *Dalbergia sissoo* Cultured in MS Medium Containing 0.5-3.0 mg l^{-1} BA alone or in Combination of 0.05 mg l^{-1} NAA. (Vertical bars represent standard error of the means)

medium containing half-strength macronutrients and augmented with 1.0 mg l^{-1} BA plus 0.05 mg l^{-1} NAA. For shoot elongation, buds along with the explants were transferred to MS medium containing the same growth regulators, but with concentrations of the macro-elements reduced to half. Buds elongated into shoots within 10-12 days of transfer to this medium. The results obtained on adventitious shoot regeneration from semi-mature cotyledon explants did not vary noticeably between *Dalbergia* species examined.

Mature Cotyledons

Mature cotyledons (control as well as the ones pre-treated with 2.0 mg l^{-1} BA) failed to respond morphogenically to a growth regulator-free MS medium. Shoot bud differentiation started only when BA was incorporated in the medium (Singh *et al.*, 2002). About 8 per cent of untreated cotyledons (control, explants without BA pre-treatment) showed adventitious bud formation on 5.0 mg l^{-1} BA supplemented medium within 25-30 days of culture and from each explant a single shoot developed. Pre-treatment of mature cotyledons for a duration of 48 h in a liquid MS medium containing 2.0 mg l^{-1} BA prior to transfer to regeneration medium was desirable as it significantly ($p>0.05$) enhanced the percentage of adventitious bud formation. Longer duration (72 h) of BA treatment inhibited shoot bud formation. Shoot buds were differentiated from mature cotyledons (pre-treated with 2.0 mg l^{-1} BA for 48 h) on MS medium containing 0.5 -5.0 mg l^{-1} BA (Figure 2.3). The cotyledons enlarged within 8-10 days of inoculation on a BA-supplemented medium and this was followed by thickening of the proximal end from which shoot bud differentiation started. Among the various concentrations of BA evaluated for shoot bud differentiation, 5.0 mg l^{-1} was found most effective and shoot differentiation occurred in 45 per cent of the pre-treated cotyledons in 20-22 days of culture (Figure 2.3). Each explant produced a mean number of 3.8 shoots averaging 4.2 cm in 35-40 days.

Callus Establishment and Shoot Organogenesis from Callus Cultures

Initiation and Proliferation of Callus Cultures

In all three species of *Dalbergia*, the juvenile explants from seedlings were proved to be more suitable for callus initiation than the explants from the mature trees. Among seedling explants the

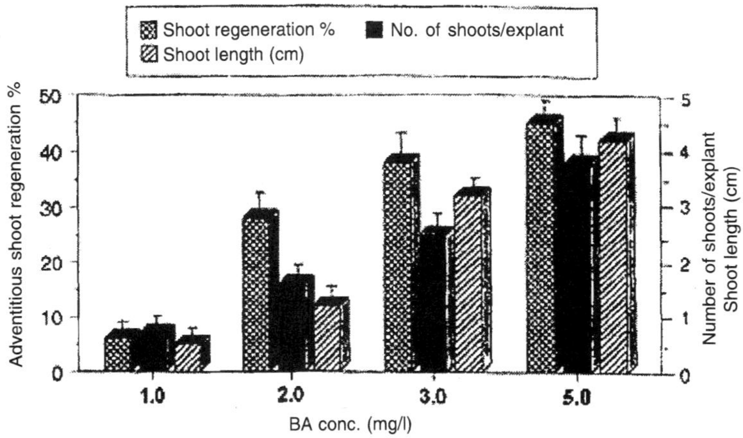

Figure 2.3: Influence of Different BA Concentrations (1.0–5.0 mg l^{-1}) in MS Medium on Adventitious Shoot Regeneration Percentage, Average Number of Shoots per Explant and Shoot Length from nature Cotyledons of *Dalbergia sissoo*. (Explants were pre-treated for 48 h in 2.0 mg l^{-1} BA; Vertical bars represent standard error of the means).

hypocotyl or cotyledonary segments were most responsive. Of the different concentrations and combinations of growth regulators tested 2.0 mg l^{-1} NAA plus 0.5 mg l^{-1} BA was found to be most effective for inducing callus formation regardless of *Dalbergia* species studied (Dwari, 1997). Of the four different basal media tested MS was identified as most efficacious for inducing callus formation in all the three species of *Dalbergia* (Figure 2.4).

In *Dalbergia lanceolaria* within 10 days of inoculation a maximum of about 97 per cent and 85 per cent callusing was induced in from the cotyledonary and hypocotyl explants respectively in MS medium supplemented with 2.0 mg l^{-1} NAA and 0.5 mg l^{-1} BA (Dwari and Chand, 1996). With the same optimal combination and concentration of phytohormones WPM gave rise to moderately growing callus in the frequencies of 52 per cent and 45 per cent from cotyledon and hypocotyl explants respectively. About 30 per cent and 25 per cent callusing was obtained in B_5 medium and 20 and 14 per cent in BM from the aforesaid explants respectively using the same optimal growth regulator combination and concentration.

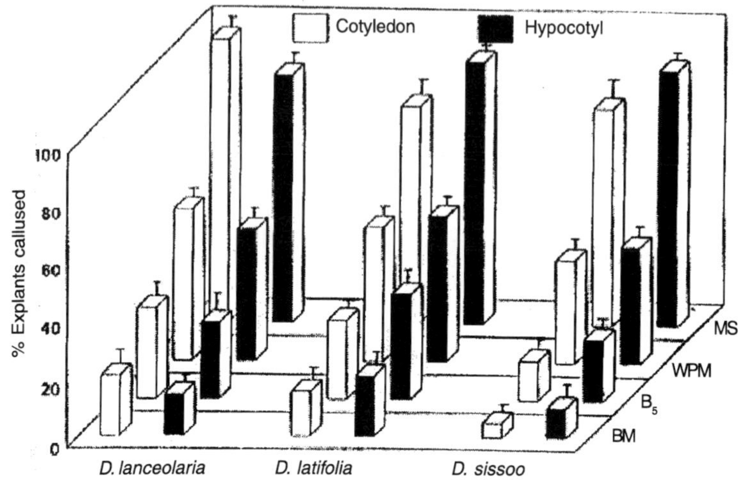

Figure 2.4: Influence of Different Basal Media on Percentage Callus Formation in Cotyledon and Hypocotyl Explants of *Dalbergia* species. (Vertical bars represent standard error of the means).

In *Dalbergia latifolia,* 90 per cent and 75 per cent callusing response was obtained in MS supplemented with 2.0 mg l^{-1} NAA and 0.5 mg l^{-1} BA from hypocotyl and cotyledonary explants respectively within 10 days of inoculation. The callus formation was about 50 per cent and 46 per cent from hypocotyl and cotyledonary explants respectively in WPM using the same concentration of NAA and BA. In B_5 medium callus formation was slow and about 35 per cent of hypocotyl and 26 per cent of cotyledon segments showed callus initiation. BM yielded most infrequent callusing in the order of 20 per cent and 15 per cent respectively using the same concentration of NAA and BA.

In *Dalbergia sissoo,* within 10 days of inoculation the frequency of callus formation was 88 per cent and 75 per cent from hypocotyl and cotyledonary explants respectively in MS medium supplemented with NAA (2.0 mg l^{-1}) and BA (0.5 mg l^{-1}). The formation of callus was about 40 per cent from hypocotyl and 35 per cent from cotyledon segments in WPM supplemented with the same concentration of NAA and BA. In B_5 medium the callus formation was very slow and occurred at a much lower frequency; about 20 per cent of hypocotyl and 13 per cent of cotyledon segments showed callus initiation. BM fortified with the above NAA-BA combination yielded poor callusing in the order of about 10 per cent from hypocotyl and 5 per cent from cotyledonary explants.

The rate of callus growth was determined in terms of gain in fresh weight and dry weight with culture passage. In *D. lanceolaria*, data on dry weight and fresh weight of cotyledon derived callus revealed an initial lag phase of growth. The lag phase was succeeded by a sudden exponential increase in the callus growth within 2nd and to 4th wk of subculture. Then there was a decline in the subsequent weeks and ultimately the calli turned brown and necrotic after the 6th week (Dwari and Chand, 1996). In *Dalbergia latifolia* and *D. sissoo* data on dry weight and fresh weight of hypocotyl derived callus showed the similar trend of growth as in *D. lanceolaria* (Dwari, 1997).

De Novo Shoot Organogenesis in Callus Cultures

In *D. lanceolaria*, the maximum shoot regeneration was elicited from cotyledon derived callus (*ca.* 97 per cent) followed by that of hypocotyl (*ca.* 70 per cent), epicotyl (*ca.* 45 per cent), primary root (*ca.* 4 per cent) and leaflet (0 per cent) in the descending order (Dwari and Chand, 1996). Of the several combinations and concentrations of auxins (IAA, NAA) and cytokinin (BA) tested in MS medium, NAA (0.5 mg l^{-1}) plus BA (2.0 mg l^{-1}) was found to be the most effective growth regulator combination for eliciting shoot regeneration from calli. There were about 12-15 shoots from hypocotyl derived callus and 18-20 shoots from callus derived from cotyledon segments. The regenerated shoots grew well and attained a height of 4-6 cm within 6-7 wk in the same medium. BA alone at all tested concentrations induced inferior frequencies of shoot regeneration from calli derived from all seedling explants. The highest frequency of shoot regeneration was about 25 per cent and 23 per cent from cotyledon and hypocotyl-derived callus respectively in the above medium with BA (2.5 mg l^{-1}) alone. It was interesting to note that there was no marked morphological changes in the callus during shoot regeneration. The callus cultures never developed green patches nor turned compact or nodular in consistency, instead maintained the original creamy white colour and friable nature throughout organogenesis.

In *D. latifolia* the maximum shoot regeneration was elicited from hypocotyl derived callus (*ca.* 76 per cent) followed by that of cotyledon (*ca.* 70 per cent), epicotyl (*ca.* 42 per cent), primary root (*ca.* 5 per cent) and leaflet (0 per cent) in the descending order (Dwari, 1997). The combination of NAA (0.5 mg l^{-1}) plus BA (2.5 mg l^{-1}) was found to be the most suitable phytohormone combination for eliciting shoot regeneration from calli. Each hypocotyl derived callus gave rise to 4-6 shoots within 4-6 wk of subculture, while only 2-4 shoots developed from callus derived from cotyledons.

In D. *sissoo* the hypocotyl derived calli were most responsive to shoot regeneration followed by those obtained from cotyledon, epicotyl and primary root in descending order. The frequency of shoot regeneration was about 30 per cent, 4 per cent and 0 per cent from epicotyl, primary root and leaflet callus respectively in the above medium (Dwari, 1997). MS medium fortified with a low concentration of NAA (0.5 mg l^{-1}) and a high concentration of BA (3.0 mg l^{-1}) was most suitable for eliciting regeneration of shoots from calli within 4-6 wk. Prior to shoot regeneration callus growth was very fast. Subsequently calli became slow-growing and turned hard, compact, green and nodular within 2-3 wk of subculture. The frequency of shoot regeneration was about 70 per cent from callus culture derived from hypocotyl and about 61 per cent from calli derived from cotyledon. There were about 3-5 shoots from a hypocotyl derived callus and 2-3 shoots per callus derived from cotyledon segments. The regenerated shoots grew well and attained a height of 3-5 cm within 7-8 wk.

Influence of Culture Age on Morphogenic Response

The morphogenic response was correlated with culture passage by plotting percentage of shoot regeneration against the culture period (Figure 2.5). In *D. lanceolaria* the shoot regeneration was not significant in the first two weeks following inoculation. This response was maximum in the 5th to 6th week and then declined sharply during the subsequent weeks. In *D. latifolia* and *D. sissoo* shoots started to develop after the third week of subculture. Shoot regeneration was maximum in the 5th and 6th week and thereafter it dropped.

The shoot regeneration potential of the calli of different age in different species was noted. There was an inverse relationship between the callus age and differentiation in all the three species of

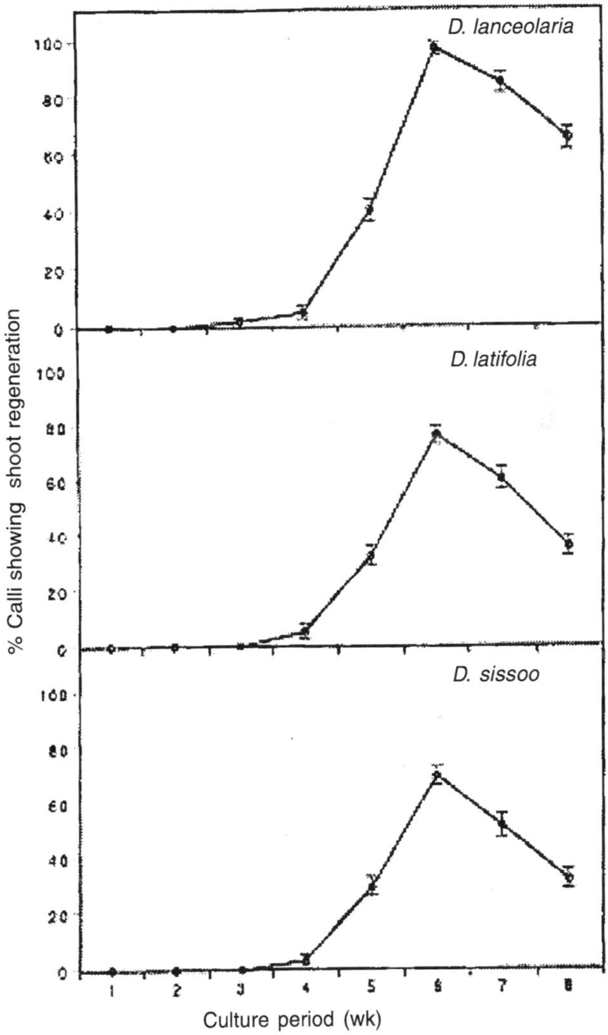

Figure 2.5: Shoot Regeneration from Two Week-Old Calli Derived from Cotyledon (*D. lanceolaria*) or Hypocotyl (*D. latifolia* and *D. sissoo*) During Different Culture Periods in the Respective Regeneration Medium. (Vertical bars represent standard error of the means).

Dalbergia studied (Dwari, 1997). The younger the calli, the better was the morphogenic performance (Figure 2.6). Maximum response was elicited from 2 wk–old calli, while regeneration was poor from

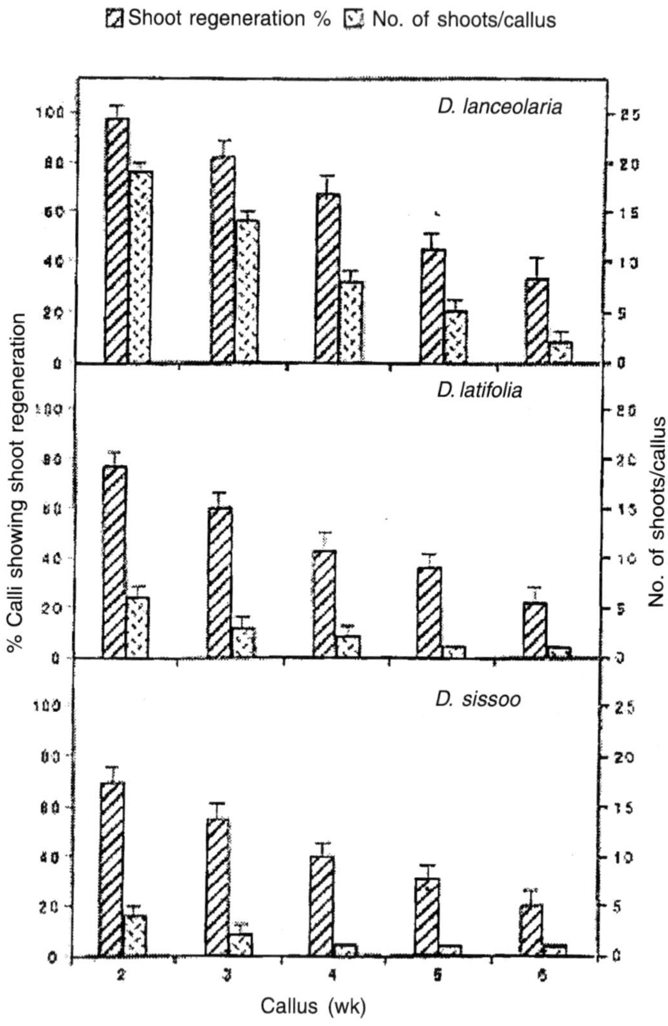

Figure 2.6: Morphogenic Potential of Callus from Cotyledon (*D. lanceolaria*) or Hypocotyl (*D. latifolia* and *D. sissoo*) Placed on the Respective Shoot Regeneration Medium Following Different Culture Passage of Growth (2-6 Week) in the Callus Proliferation Medium. (Vertical bars represent standard error of the means).

the calli which were 6-7 wk old. Calli cultured in the proliferation medium for more than 8 wk were unable to form shoots regardless of subsequent culture period in the regeneration medium.

Shoot Organogenesis in Callus-derived Cell Suspension Cultures

Establishment of Cell Suspension Cultures

Hypocotyl-derived proliferating calli from the second subculture passage were suitable for initiating cell suspension cultures in all the three *Dalbergia* species. Of the three basal nutrient media (MS, B_5, UM [Uchimiya and Murashige, 1974]) evaluated, optimum proliferation of cells were observed in MS regardless of species. Newly initiated, granular cell suspension cultures were composed of isodiametric, thin-walled and richly cytoplasmic cells. By sieving the larger cell clumps during each subculture, it was possible to establish a cell suspension composed of only finer cell clumps within 2-3 months (Pradhan *et al.*, 1998c; Pattnaik *et al.*, 2000). The viability of the newly initiated suspension cultures (second passage) varied between 92 per cent–95 per cent for different *Dalbergia* species based on FDA staining. Viability decreased with the age of the culture such that the cells harvested from the tenth culture passage had a viability ranging 35-40 per cent in all the species.

The growth curve based on settled cell volume (SCV, per cent) of the cell suspension culture showed that there was an initial lag phage of 2-3 days (*D. sissoo*) or 3-4 days (*D. lanceolaria* and *D. latifolia*), which was followed by the exponential growth phase. In *D. sissoo* exponential growth occurred between day 4-10 in MS, day 4-8 in B_5 and day 4-6 in UM, but in *D. lanceolaria* as well as *D. latifolia* exponential growth took place between day 6-12 in MS and B_5 and day 6-9 in UM. However, in all the species cell proliferation was the highest in MS. There was no cell proliferation after 12 days for *D. sissoo* or after 15 days for *D. lanceolaria* and *D. latifolia*.

Growth of the cell suspension culture was influenced by initial cell density of the inoculum (Pradhan, 2000; Pattnaik *et al.*, 2000). For *D. lanceolaria* as well as *D. sissoo* a 1 per cent inoculum cell density was the best for establishment of suspension cultures with an optimal growth of cells. On the other hand, a 2 per cent inoculum cell density was determined to be the best for establishment of suspension cultures of *D. latifolia* with an optimal proliferation of cells.

Microcolonies (15-20 cells) harvested from suspension culture proliferated into larger colonies (macrocolonies, 2-3 mm diam) within 10-15 days on MS medium with NAA (2.0 mg l^{-1}), BA (0.5 mg l^{-1}) and 10 per cent coconut water (Figure 2.9b). Plating efficiency of cell clumps harvested during exponential growth phase was the maximum irrespective of species (Figure 2.7). Plating efficiency also varied with the type of gelling agents used for embedding the cells. Of the three different gel matrices employed for immobilizing the microcolonies the highest plating efficiency (93 per cent in *D. sissoo*, 92 per cent in *D. lanceolaria* and 89 per cent in *D. latifolia*) was recorded with phytagel (Figure 2.8). Cells harvested at their exponential growth phase (4-10 days) had the highest plating efficiency which declined with cells harvested either earlier (2-4 days) or later (10-16 days).

Shoot Organogenesis in Cell Suspension-derived Callus

Colonies were developed into calli which turned dark-green, compact and hard within 3-4 weeks of transfer to the regeneration medium. Shoot buds were differentiated from these calli after the 4th week in *D. sissoo* and after the 5th wk in *D. lanceolaria* as well as *D. latifolia*. In all the species shoot bud differentiation continued until the end of 8th wk. High frequency shoot organogenesis *i.e.* 71 per

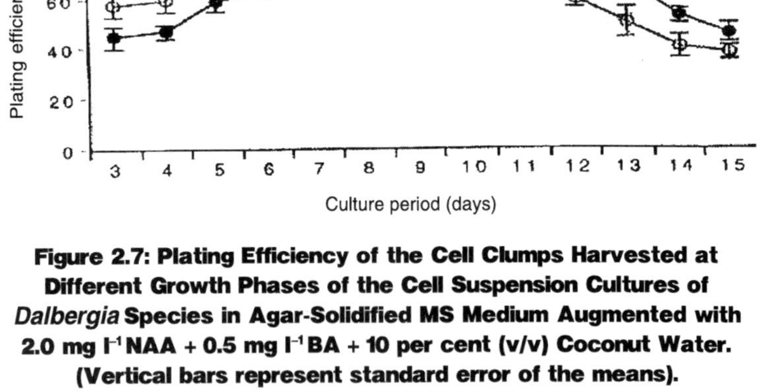

Figure 2.7: Plating Efficiency of the Cell Clumps Harvested at Different Growth Phases of the Cell Suspension Cultures of *Dalbergia* Species in Agar-Solidified MS Medium Augmented with 2.0 mg l^{-1} NAA + 0.5 mg l^{-1} BA + 10 per cent (v/v) Coconut Water. (Vertical bars represent standard error of the means).

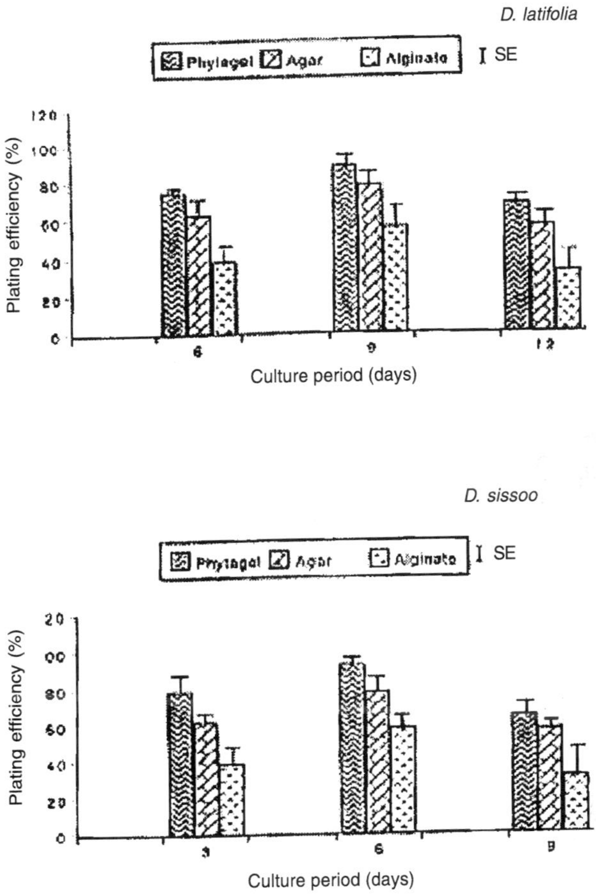

Figure 2.8: Influence of Different Gelling Agents on the Plating Efficiency of the Cell Clumps Harvested at Different Culture Periods from the Exponential Growth Phase of the Cell Suspensions of *Dalbergia* species. (a) *D. latifolia* (b) *D. sissoo*. (Vertical bars represent standard error of the means).

cent or 79 per cent in *D. sissoo* and *D. lanceolaria*/*D. latifolia* respectively coupled with maximum shoot bud formation was observed in a medium containing 3.0 mg l^{-1} BA and 0.5 mg l^{-1} NAA. Suspension-derived callus of *D. sissoo* produced a mean number of 5.9 shoots averaging 4.5 cm in length in 8-10 wk whereas in *D. lanceolaria* 5.8 shoots produced averaging 4.4 cm and in *D. latifolia* 5.6 shoots

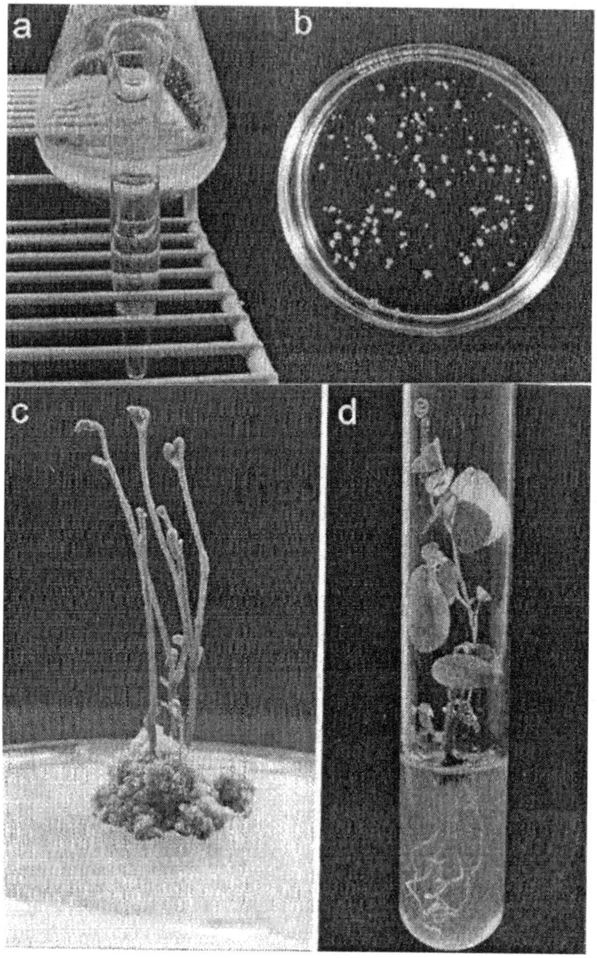

Figure 2.9: Plant Regeneration from Cell Suspension Culture of *D. latifolia*

(a) Measurement of growth (settled cell volume, SCV per cent) of cell suspension culture using a culture flask (250 ml) fitted with a graduated side arm; (b) Development of macrocolonies from phytagel (1.2 g l^{-1})-embedded microcolonies 15 days after culture in MS + 2.0 mg l^{-1} NAA + 0.5 mg l^{-1} BA; (c) Elongated multiple shoots following regeneration from a cell suspension-derived callus in MS + 3.0 mg l^{-1} BA + 0.5 mg l^{-1} NAA on day 70 of culture; (d) A rooted shoot in □ MS + 1.0 mg l^{-1} IAA + 1.0 mg l^{-1} IBA + 1.0 mg l^{-1} IPA on day 30 of culture.

produced with average height of 4.2 cm (Figure 2.9c; Pradhan *et al.*, 1998; Patnaik *et al.*, 2000). Shoot bud differentiation was also induced on a medium containing BA alone in *D. sissoo* or BA combined with IAA in *D. lanceolaria* and *D. latifolia*, but in both cases the regeneration frequency was much lower and number of shoot buds per callus were fewer. At higher concentrations of BA the frequency of shoot-bud differentiation declined and shoots that developed from the calli became fasciated.

The morphogenic competence of the suspension-derived calli was influenced by the age of the liquid culture (Pradhan, 2000). It was observed that calli derived from the second, third and fourth subculture passage retained a high regeneration potential with respect to frequency of shoot bud regeneration as well as number of shoots per callus. However, calli obtained from suspension older than the fourth passage showed a gradual decline in the regeneration potential.

Root Induction in Shoots and Field Establishment of *in vitro* Regenerated Plants

Rooting of regenerated shoots occurred within 18-21 days in all the three species of *Dalbergia* only when the culture medium used contained an auxin supplement. White's basal medium (1963) fortified with IPA (1.0 mg l^{-1}) was superior to the MS formulation with a similar phytohormone additive (Dwari and Chand, 1996). Nevertheless, a comparable rhizogenesis occurred in half-strength MS medium only when supplemented with 1.0-2.0 mg l^{-1} IBA. In 30 days, there were about 3-4 roots formed per shoot, roots averaging 4.2 cm in length. The percentage of rooting was markedly enhanced by supplementing the medium with a combination of IAA, IBA and IPA each at 1.0 mg l^{-1} (Pradhan *et al.*, 1998a-c; Patnaik *et al.*, 2000). In 7 days almost all shoots of *D. lanceolaria* rooted in this medium whereas in case of *D. sissoo* 91 per cent of shoots rooted within 6-8 days and in *D. latifolia* 72 per cent rooted within 8-10 days. The roots elongated accompanied by the development of a few new roots following transfer to a medium without growth regulators. In 30 days, about 5-6 roots per shoot were developed with a mean root length recorded as 5.3 cm. Other auxins (IAA or NAA) when used singularly were not effective in inducing root formation regardless of concentrations of basal medium.

The rooted plantlets (Figure 2.9d) were successfully hardened off in the plant growth chamber for 3 wk followed by another 2 wk in the shade and eventually established outdoors under full sun. The percentage survival of *in vitro* regenerated plants ranged 68-70 per cent after transfer to vermi-compost and 60-80 per cent following transplantation in natural soil.

Discussion

The successful establishment of *in vitro* cultures with optimum growth of plant cells is largely dependent on three factors which include (i) choice of explant type (ii) composition of nutrient medium and (iii) the physical environment (Thorpe and Patel, 1984). Of these, the composition of the culture medium is the most important parameter which must be optimized in order to initiate *in vitro* cultures and to achieve plant regeneration from such cultures. Among the four different kinds of basal nutrient media formulations tested in the present studies MS was found to be the most suitable for all types of *in vitro* manipulations attempted and described in this paper. These manipulations resulted in successful plant regeneration of three economically important tree legume species of *Dalbergia* via (i) axillary shoot proliferation from cotyledonary nodes (ii) adventitious shoot organogenesis in semi-mature and mature cotyledons (iii) shoot organogenesis in callus and cell suspension cultures. Although MS has been proven useful in several plant species, present results differed from that in some other tree legume species including *Albizzia lebbeck* (Gharyal and Maheswari, 1983), *Acacia auriculiformis* (Mittal *et al.*, 1989), *Sesbania* species (Shankar and Mohan Ram, 1990) and *Pterocarpus santalinus* (Lakshmi Sita *et al.*, 1992) in which B_5 medium was most effective.

BA-induced axillary shoot proliferation in the cotyledonary nodes of seedlings have been reported in several tree species such as *Alnus glutinosa* (L.) Gaertn. (Perinet and Lalonde, 1983), *Wrightia tinctoria* R.Br. (Purohit and Kukda, 1994), *Melissa officinalis* L. (Tavares *et al.*, 1996), *Sterculia urens* Roxb. (Purohit and Dave, 1996), *Murraya koenigii* (L.) Spreng. (Bhuyan *et al.*, 1997), *Achras sapota* (Purohit and Singhvi, 1998) and *Punica granatum* (Naik *et al.*, 2000). The frequency of shoot proliferation declined markedly at higher concentrations of cytokinins regardless of *Dalbergia* species studied. Moreover, the shoots that developed on these media failed to elongate. These results were consistent with that reported in pomegranate (Naik *et al.*, 2000).

Thidiazuron has been reported to induce axillary as well as adventitious shoots in many tree species especially recalcitrant ones (Huetteman and Preece, 1993). TDZ at 0.01-0.1 mg l^{-1} was used in the present study to induce axillary shoots in the cotyledonary nodes of *Dalbergia* species. However, it was the least effective of all the cytokinins tested for inducing shoot proliferation. Although multiple shoots were induced in a TDZ-containing medium, the shoots failed to elongate and were often fasciated. The low frequency shoot formation or stunted shoots or the fasciation of shoots on TDZ-supplemented medium has been reported for several tree species such as *Malus* spp. (van Nieuwkerk *et al.*, 1986), *Populus* sp. (Russel and McCown, 1986), *Rhododendron* spp. (Preece and Imel, 1991) and *Punica granatum* L. (Naik *et al.*, 1999). Inhibition of shoot elongation may be due to the high cytokinin activity of TDZ, whereas the presence of a phenyl group in TDZ may be the possible cause of shoot-bud fasciation (Huetteman and Preece, 1993). Unlike the adenine-type cytokinins (BA, Kin, 2iP, ZR), TDZ is more effective at lower concentrations and, therefore, direct comparison of TDZ with each of these cytokinins at the same concentration is not possible.

Data presented showed that semi-mature and mature cotyledons of *Dalbergia* species were capable of producing adventitious shoots. However, the requirement for plant growth regulators varied for shoot bud differentiation and rooting of *in vitro* formed shoots. The mature cotyledons required only BA, whereas the semi-mature ones preferred a combination of BA and NAA in the medium for shoot bud differentiation. This differential requirement of growth regulators could be due to the difference in the developmental stage and endogenous level of plant growth hormones between semi-mature and mature cotyledons. Morphological and biochemical characteristics of cotyledons change with age and these changes can affect cytokinin uptake and competency of cells to initiate shoot buds (Bornman, 1983). Pre-treatment of mature cotyledons with 2.0 mg l^{-1} BA for 48 h markedly enhanced the percentage response for adventitious shoot formation. The promotive effect of pulse treatment with BA on adventitious shoot regeneration has been reported in Douglas-fir cotyledons (Goldfarb *et al.*, 1991). According to Krul and Myerson (1980) plant regeneration through direct adventitious shoot organogenesis is genetically a more stable strategy compared to that through an intermediate de-differentiated state of culture.

In the present investigation best callusing was noticed in MS fortified with 2.0 mg l^{-1} NAA and 0.5 mg l^{-1} BA regardless of *Dalbergia* species examined. The highest percentage of shoot regeneration with maximum number of shoots per callus were also achieved in the same basal medium but containing NAA-BA combinations with concentrations of BA 4-6 fold higher than that of NAA. These results have shown that a relatively high concentration of auxin combined with a lower concentration of cytokinin in the medium promoted abundant cell proliferation and de-differentiation resulting in the formation of callus. On the contrary, a relatively low auxin plus a higher cytokinin content in the medium triggered shoot initiation from callus cultures. These observations corroborate the significance of auxin-cytokinin balance to regulate callus induction and shoot differentiation in plant tissue cultures as hypothesized by Skoog and Miller (1957).

It was interesting to note that there was a differential response among the seedling explants to callus initiation and shoot regeneration. In *Dalbergia lanceolaria* cotyledonary segments were more amenable than other seedling expants in terms of callusing and shoot differentiation whereas in *D. latifolia* and *D. sissoo* hypocotyl segments were most responsive. Explant-specific morphogenic responsiveness was detected even among different types of explants from the same plant and tree legumes were no exception (Kapoor and Gupta, 1986; Lakshmana Rao and De, 1987). Certain enzymatic differences were shown to occur between callus cultures derived from different tissues of the same plant suggesting that perhaps all tissue types fail to de-differentiate and/or differentiate to the same extent (Lindsey and Yeoman, 1985).

It was also observed that the frequency of regeneration as well as the number of shoots per callus increased with increasing BA concentration in combination with the same concentration of NAA and after reaching a maximum value it declined regardless of further increments in BA content. This was true for all the three *Dalbergia* species examined. Similar findings were reported for other tree legumes such as *Leucaena leucocephala* (Goyal *et al.*, 1985) and *Sesbania bispinosa* (Sinha and Mallick, 1991). The nature of calli and the frequency of shoot regeneration from callus cultures were found to vary with *Dalbergia* species. Similar findings have been reported in other tree species including *Malus* (Zimmerman and Broome, 1981) and *Rhododendron* (McCown and Lloyd, 1983). Such observations were indicative of an inherent genotypic variation among various species within the same genus.

There was an inverse relationship between callus age and differentiation; younger calli were superior with respect to morphogenic performance. Similar observations were reported for other tree legume species such as *Pterocarpus santalinus* (Patri *et al.*, 1988), *Sesbania grandiflora* (Shankar and Mohan Ram, 1990) and *S. bispinosa* (Sinha and Mallick, 1992) as well as for other forest tree species including *Morus indica* (Sahoo *et al.*, 1997). The retrogression in the morphogenic potential of calli might be due to the accumulation of inhibitory substances (Halperin, 1986), decline in metabolism, transport and interaction between growth regulators (Hansen *et al.*, 1987; Van Staden and Moony, 1987; Palni *et al.*, 1988) or increase in the number of polyploid and aneuploid cells in the callus because of prolonged subculture (Nehra *et al.*, 1990).

A complete and reliable protocol is presented for establishment of a fast-growing cell-suspension culture using seedling explants of *Dalbergia* species and high frequency shoot organogenesis from callus derived from such cultures. It has been demonstrated that highly regenerable cell-suspension cultures can be established using seedling explants of all the three species of *Dalbergia*. Parameters that influenced the establishment of cell-suspension cultures and plant regeneration from such cultures were optimized. Plant regeneration from cell-suspension cultures established using cambial tissue of a mature tree has been reported in *D. sissoo* (Kumar *et al.*, 1991); but the maximum shoot-regeneration frequency never exceeded 60 per cent on a medium containing 1.0 mg l^{-1} BA alone. Besides, this report is silent with respect to various aspects such as the growth pattern and viability of the cell-suspension culture, the initial inoculum cell density required for optimum growth of the cultures, and the plating efficiency of the embedded microcolonies, which are considered critical for establishment of such cultures. As shown in the present studies, cell suspension cultures derived from juvenile tissues (hypocotyl) were more competent with regard to their morphogenic potential compared to those of cambial origin. In comparison with the alternative *in vitro* propagation methods previously reported, the present protocol based on plant regeneration from cell-suspension cultures will ensure large-scale production of this important forest tree, while offering a considerable opportunity for recovering useful somaclonal variants from the existing gene pool.

Advantages of cell-suspension cultures over stationary cultures are numerous. As cells in suspension are amenable to the influence of exogenous chemicals such as mutagens, such cultures can be used effectively for the isolation of mutant cell lines or somaclonal variants. Moreover, these cultures provide a viable alternative for commercial production of a wide array of plant metabolites (Bhojwani and Razdan, 1996). Suspension culture is the most suitable mode for studying the physiological, biochemical and molecular basis of organ differentiation (Ozawa *et al.*, 1996). However, to date, establishment of regenerable cell-suspension cultures has been reported only in a few woody species, including *Actinidia chinensis* (Suezawa *et al.*, 1988) and *Populus alba* (Park and Son, 1988).

As compared to earlier methods, the establishment of fast-growing highly regenerable cell-suspension cultures would be advantageous for massive propagation of *Dalbergia* species by exploiting the cellular totipotency of higher plants. Such cultures, particularly when derived from seedling explant sources, offer the potential for creating novel somaclonal variations among the regenerated plant population and thus provide an additional tool for cellular genetic manipulation aimed at improvement of these three important timber yielding-cum-medicinal forest tree species of the genus *Dalbergia*.

The enhanced rooting ability of *in vitro*-regenerated shoots elicited by the combination of three different auxin supplements might be due to a synergistic effect. Combined presence of these three auxins was reported to have favoured root induction in several tree species including *Tectona grandis* (Bajaj, 1986), *Cinnamomum zeylanicum* (Ravishankar Rai and Jagadish Chandra, 1987), *Morus australis* Poir. (Pattnaik *et al.*, 1996), *M. cathayana* Hemsl., *M. lhou* Koiz. and *M. serrata* Roxb. (Pattnaik and Chand, 1997).

In essence, the protocols embodied in this paper for direct callus-free plant regeneration either through *in vitro* multiple shoot proliferation in cotyledonary nodes or via adventitious shoot formation in cotyledon explants lacking pre-existing meristem could be useful for rapid mass production of true-to-type plants needed for large scale plantation of these economically important forest tree legumes of the genus *Dalbergia.* On the other hand, efficient plant regeneration by high frequency shoot organogenesis via an intermediate de-differentiated state of culture such as callus or callus-derived cell suspensions ushers a possibility of genetic improvement through *in vitro* selection for generation of novel somaclonal variants and mutants, somatic hybrids and cybrids as well as transgenic plants, unachievable by conventional tree breeding.

Acknowledgement

The financial assistance from the Department of Biotechnology, Ministry of Science and Technology, Government of India, through a major R&D Project is gratefully acknowledged.

References

Anonymous (1970). The Wealth of India: A dictionary of Indian raw materials and industrial products, Vol 3 (D-E) CSIR, New Delhi. PP. 3-12.

Bajaj YSP (1986) Biotechnology of tree improvement for rapid propagation and biomass energy production. In: Biotechnology in agriculture and forestry. Vol. I, Trees I.Ed. Bajaj YSP. Springer-Verlag, Berlin, PP. 1-23.

Bhojwani SS and Razdan MK (1996) Cell culture. In: Plant Tissue Culture: Theory and Practice, Revised edn. Eds. Bhojwani SS and Razdan, MK. Elsevier, Amsterdam, PP. 63-93.

Bhuyan AK, Pattnaik S and Chand PK (1997) Micropropagation of curry leaf tree [*Murraya koenigii* (L.) Spreng.] by axillary proliferation using intact seedlings. Plant Cell Rep. 16: 779-782.

Blayes DF (1966) Interaction of kinetin and various inhibitors in the growth of soybean tissue. Physiol. Plant. 19: 748-753.

Bornman CH (1983) Possibilities and constraints in the regeneration of trees from cotyledonary needles of *Picea abies. in vitro.* Physiol. Plant. 57: 5-16.

Chopra RN, Nayar SL and Chopra IC (1956) Glossary of Indian medicinal plants. Council of Scientific and Industrial Research, New Delhi, P.90.

Das P, Samantaray S, Roberts AV and Rout GR (1997) *In vitro* somatic embryogenesis of *Dalbergia sissoo* Roxb., a multipurpose timber yielding tree. Plant Cell Rep. 16: 578-582.

Datta SK and Datta K (1983) Auxin induced regeneration of forest tree *Dalbergia sissoo* Roxb. through tissue culture. Curr. Sci. 52: 434-436.

Datta SK, Datta K and Pramanik T (1983) *In vitro* clonal multiplication of mature trees of *Dalbergia sissoo* Roxb. Plant Cell Tiss. Org. Cult. 2: 15-20.

Dwara S, Sharma DR and Choudhury JB (1984) Clonal propagation of *Dalbergia sissoo* Roxb. through tissue culture. Curr. Sci. 53: 807-809.

Dwari M and Chand PK (1996) Evaluation of explants, growth regulators and culture passage for enhanced callus induction, proliferation and plant regeneration in the tree legume *Dalbergia lanceolaria* L. Phytomorphol. 46(2): 123-131.

Dwari M (1997) Changes in levels of some metabolites and enzymes during callus growth and organogenesis in *Dalbergia* species. Ph. D Thesis, Utkal University, Bhubaneswar, India.

Gamborg OL, Miller RA and Ojima K (1968) Nutrient requirement of suspension cultures of soybean root cells. Expt. Cell Res. 50: 151-158.

Gharyal PK and Maheswari SC (1990) *In vitro* differentiation of plantlets from tissue cultures of *Albizzia lebbeck* L. Plant Cell Tissue Org. Cult. 2: 49-53.

Goldfarb B, Howe GT, Bailey LM, Strauss SH and Zaerr JB (1991) A liquid cytokinin pulse induces adventitious shoot formation from Douglas-fir cotyledons. Plant Cell Rep. 10: 156-160.

Goyal Y, Bingham RL and Felker P (1985) Propagation of tropical tree, *Leucaena leucocephala* K_{67} by *in vitro* bud culture. Plant Cell Tiss. Org. Cult. 3-10.

Gulati A and Jaiwal PK (1996) Micropropagation of *Dalbergia sissoo* Roxb. from nodal explants of mature trees. Biol. Plant. 38: 169-175.

Halperin W (1986) Attainment and retention of morphogenetic capacity *in vitro*. In: Cell culture and Somatic Cell Genetics of

Plants, Ed. Vasil IK, Vol. 3, Academic Press, Orlando, Florida PP. 3-47.

Hansen CE, Meins F and Aebi R (1987) Hormonal regulation of zeatin-riboside accumulation by tobacco cells. Planta, 172: 520-525.

Huetteman CA and Preece JE (1993) Thidiazuron: a potent cytokinin for woody plant tissue culture. Plant Cell Tiss. Org. Cult. 33: 105-119.

Kapoor S and Gupta S (1986) Rapid *in vitro* differentiation of *Sesbania bispinosa* plants: A leguminous shrub. Plant Cell Tiss Org Cult. 7: 263-268.

Krul WR and Myerson J (1980) *In vitro* propagation of grape. In: Proc. Conf. Nursery Production of Fruit Plant through Tissue Culture-Application and Feasibility, Ed. Zimmerman RH. Agric. Res. Sci. Educ. Admin. USDA, Beltsville, PP. 35-43.

Kumar A, Tandon P and Sharma A (1991) Morphogenic responses of cultured cells of cambial origin of a mature tree *Dalbergia sissoo* Roxb. Plant Cell Rep. 9: 703-706.

Lakshmana Rao PV and De DN (1987) Tissue culture propagation of tree legume *Albizzia lebbeck* (L.) Benth. Plant Sci. 51: 263-267.

Lakshmi Sita G, Sreenath KS and Sujatha S (1992) Plantlet production from shoot tip cultures of red sandalwood (*Pterocarpus santalinus* L.). Curr. Sci. 62: 532-535.

Larkin PJ (1976) Purification and viability determination of plant protoplasts. Planta 128: 213-216.

Lindsey K and Yeoman MM (1985) Dynamics of plant cell cultures. In: Cell culture and Somatic Cell Genetics of Plants, Ed. Vasil IK, Vol.2, Academic Press, Orlando, PP. 61-101.

Lloyd GB and McCown BH (1981) Commercially feasible micropropagation of mountain laurel, *Kalmia latifolia*, by use of shoot tip culture. Proc. Int. Plant Prop. Soc. 30: 421- 427.

McCown BH (2000) Recalcitrance of woody and herbaceous perennial plants; Dealing with genetic predetermination. *In vitro* Cell. Dev. Biol. Plant. 36: 149-154.

McCown BH and Lloyd GB (1983) A survey of the response of *Rhododendron* to *in vitro* culture. Plant Cell Tiss. Org. Cult. 2: 77-85.

Mittal A, Agarwal R and Gupta SC (1989) *In vitro* development of plantlets from axillary buds of *Acacia auriculiformis*, a leguminous tree. Plant Cell Tissue Org. Cult. 19: 65-70.

Murashige T and Skoog F (1962) A revised medium for rapid growth and bioassays with tobacco tissue cultures. Physiol. Plant., 15: 473-497.

Naik SK, Pattnaik S and Chand PK (1999) *In vitro* propagation of pomegranate (*Punica granatum* L. cv. Ganesh) through axillary shoot proliferation from nodal segments of mature tree. Scient. Hort. 79: 175-183.

Naik SK, Pattnaik S and Chand PK (2000) High frequency axillary shoot proliferation and plant regeneration from cotyledonary nodes of pomegranate (*Punica granatum* L.) Scient. Hort. 85: 261-270.

Nayar MP (1996) Hotspots of endemic plants of India, Nepal and Bhutan, Thiruvananthapuram, Kerala.

Nehra NS, Stushnoff C and Kartha KK (1990) Regeneration plants from immature leaf–derived callus of strawberry (*Fragaria* x *Ananassa)*. Plant Sci. 66: 119-126.

Ozawa K, Ling DH and Komamine A (1996) High frequency somatic embryogenesis from small suspension-cultured clusters of cells of interspecific hybrid of *Oryza*. Plant Cell Tiss. Org. Cult. 46: 157-159.

Palni LMS, Burch L and Horgan R (1988) The effect of auxin concentration on cytokinin stability and metabolism. Planta 174: 231-234.

Park YG and Son SH (1988) Regeneration of plantlets from cell suspension derived callus of white poplar (*Populus alba* L.). Plant Cell Rep. 7: 567-570.

Pattnaik S and Chand PK (1997) Rapid clonal propagation of three mulberries *Morus cathayana* Hemsl., *M. lhou* Koiz. and *M.serrata* Roxb. through *in vitro* culture of apical shoot buds and nodal explants from mature trees. Plant Cell Rep. 16: 503-508.

Pattnaik S, Pradhan C, Naik SK and Chand PK (2000) Shoot organogenesis and plant regeneration from hypocotyl–derived cell suspensions of a tree legume *Dalbergia sissoo* Roxb. *In Vitro* Cell Dev Biol–Plant 36(5): 407-411.

Pattnaik SK, Sahoo Y and Chand PK (1996) Micropropagation of a fruit tree, *Morus australis* Poir. Syn. *M. acidosa* Griff. Plant Cell Rep. 15: 841-845.

Patri S, Bhatnagar SP and Bhojwani SS (1988) Preliminary investigation on micropropagation of a leguminous timber tree: *Pterocarpus santalinus*. Phytomorphol. 38: 41-45.

Perinet P and Lalonde K (1983) *In vitro* propagation and nodulation of actinorrhizal host plant, *Alnus glutinosa* (L.) Gaertn. Plant Sci.Lett. 29: 9-17.

Pradhan C, Kar S, Pattnaik SK and Chand PK (1998a) Propagation of *Dalbergia sissoo* Roxb. through *in vitro* shoot proliferation from cotyledonary nodes. Plant Cell Rep. 18: 122-126.

Pradhan C, Pattnaik SK and Chand PK (1998b) Rapid *in vitro* propagation of East Indian Rosewood (*Dalbergia latifolia* Roxb.) through high frequency shoot proliferation from cotyledonary nodes. J Plant Biochem. Biotechnol. 7: 61-64.

Pradhan C, Pattnaik SK, Dwari M, Patnaik SN and Chand PK (1998c) Efficient plant regeneration from cell suspension derived callus of East Indian Rosewood (*Dalbergia latifolia* Roxb.). Plant Cell Rep. 18: 138-142.

Pradhan C (2000) Plant regeneration from tissue and cell culture and *Agrobacterium*-mediated genetic transformation of *Dalbergia* species. Ph.D Thesis, Utkal University, Bhubaneswar, India.

Preece JE and Imel MR (1991) Plant regeneration from leaf explants of *Rhododendron* P.J.M hybrids. Scient. Hort. 48: 159-170.

Purohit SD and Dave A (1996) Micropropagation of *Sterculia urens* Roxb. an endangered tree species. Plant Cell Rep. 15: 704-706.

Purohit SD and Kukda G (1994) *In vitro* propagation of *Wrightia tinctoria*. Biol. Plant. 36: 519-526.

Purohit SD and Singhvi A (1998) Micropropagation of *Achras sapota* through enhanced axillary branching. Scient. Hort. 76: 219-229.

Ravishankar Rai V and Jagadishchandra KS (1987) Clonal propagation of *Cinnamomum zeylanicum* Breyn by tissue culture. Plant Cell Tiss. Org. Cult. 9: 81-88.

Rao CK Geetha BL and Suresh G (2003) Red list of threatened vascular plant species of India. Director, Botanical Survey of

India, ENVIS centre for floral diversity. Botanical Survey of India. Central National Herbarium, Indian Botanical Garden, Howrah, India.

Rodgers WA and Panwar HS (1990) A biogeographical classification for conservation planning. Wildlife Institute of India, Dehradun, India.

Russel, JA and McCown, BH (1986) Thidiazuron stimulated shoot differentiation from protoplast-derived callus of *Populus*. Abs. VI Intl. Cong. Plant Tissue Culture, Amsterdam, PP. 49.

Sahoo Y, Pattnaik SK and Chand PK (1997) Plant regeneration from callus cultures of *Morus indica* L. derived from seedlings and mature plants. Scient. Hort. 69: 85-98.

Sharma S and Chandra N (1988) Organogenesis and plantlet formation *in vitro* in *Dalbergia sissoo* Roxb. J. Plant Physiol. 132: 145-147.

Shankar S and Mohan Ram HY (1990) Plantlet regeneration from tissue cultures of *Sesbania grandiflora*. Curr. Sci. 59: 39-43.

Shivarajan VV and Balachandran I (1999) Ayurvedic drugs and their plant sources. Oxford and IDBH Publishing Co. Pvt. Ltd., New Delhi, PP.9.

Singh AK, Chand S, Pattnaik S and Chand PK (2002) Adventitious shoot organogenesis and plant regeneration from cotyledons of *Dalbergia sissoo* Roxb., a timber yielding tree legume. Plant Cell Tiss. Org. Cult. 68: 203-209.

Singh NP and Chowdhery HJ (2002) Biodiversity conservation in India. In: Perspective of Plant Biodiversity, Ed. Das, AP. Department of Botany, North Bengal University, West Bengal, India, Bishen Singh Mahendra Pal Singh, Dehradun-248001, India, PP. 501-527.

Sinha RK and Mallik R. (1991) Plantlets from somatic callus tissue of the woody legume *Sesbania bispinosa* (Jacq.) W.F. Wight. Plant Cell Rep. 10: 247-250.

Skoog F and Miller CO (1957) Chemical regulation of growth and organ formation in plant tissue culture *in vitro*. Symp. Soc. Expt. Biol. 11: 118-131.

Suezawa K, Matsuta N, Omura M and Yamaki S (1988) Plantlet formation from cell suspensions of kiwi fruit (*Actinidia chinensis* Planch. var. chinensis). Scientia Hortic. 37: 123-128.

Tavares AC, Pimenta MC and Goncalves MT (1996) Micropropagation of *Melissa officinalis* L. through proliferation of axillary shoots. Plant Cell Rep. 15: 441-444.

Thorpe TA and Patel KR (1984) Clonal propagation: Adventitious buds. In: Cell Culture and Somatic Cell Genetics of Plants, Ed. Vasil, IK, Vol. 1, Academic Press, New York, PP. 49-60.

Tomar UK and Gupta SC (1988) *In vitro* plant regeneration of leguminous trees (*Albizia* spp.). Plant Cell Rep. 7: 385-388.

Trigiano RN, Geneve R, Merkle SA, Preece JE (1992) Tissue and cell cultures of woody legumes. Hort. Rev. 14: 265-332.

Uchimiya H and Murashige T (1974) Evaluation of parameters in the isolation of viable protoplasts from cultured tobacco cells. Plant Physiol. 54: 936-944.

Van Nieuwkerk JP, Zimmerman RH and Fordham I (1986) Thidiazuron stimulation of apple shoot proliferation *in vitro*. Hort Sci. 21: 516-518.

Van Staden J and Moony PA (1987) Effect of IAA on the metabolism of kinetin by *Catharanthus roseus* crown gall callus. J. Plant Physiol. 131: 297-303.

White PR (1963) Cultivation of animal and plant cells. Ronald Press. Co. N.Y. PP. 228-229.

Yoganarasimhan SN (2000) Medicinal plants of India. Interline Publishing Co., Bangalore, India, Vol.2: PP.181-182.

Zimmerman RH and Broome OC (1981) Phloroglucinol and *in vitro* rooting of apple cultivar cuttings. J. Am. Soc. Hort. Sc. 106: 648-652.

Chapter 3

Chlorophyll Fluorescence: An Easy Tool for Identification of Submergence Tolerant Rice Cultivar

D. Panda*, S.G. Sharma and R.K. Sarkar

Division of Biochemistry,
Plant Physiology and Environmental Sciences,
Central Rice Research Institute,
Cuttack – 753 006, Orissa, India

ABSTRACT

The influence of submergence on Photosystem II (PS II) activity and leaf chlorophyll content were examined in three *Indica* rice (*Oryza sativa* L.) cultivars (e.g. FR 13A, IR 42 and Sabita which are tolerant, susceptible and avoiding type respectively) under 10 days of submergence. Chlorophyll a fluorescence transients were recorded in fully dark-adapted leaves of control and submerged plants. The recorded transients were analyzed by so called JIP-test, revealed that there were large cultivars difference in response of PS II to submergence. Among them maximal fluorescence (Fm), potential quantum yield of primary photochemistry (Fv/Fm), area above the fluorescence curve between minimum (Fo) and maximum fluorescence (Fm), performance index (PI), electron transport

* Corresponding Author: E-mail: dpanda80@gmail.com; Ph: 91-9437511187

per cross section (ETo/CS) and active reaction center per cross section (RC/CS) decreased due to submergence in all the cultivars. Submergence also decreased the total chlorophyll content in the leaves. These parameters were differentially changed in the three genotypes during the progression of the submergence treatment. Different chlorophyll fluorescence parameters could clearly differentiate between susceptible and tolerant cultivars as early as 4 to 6 d of submergence when even the susceptible cultivar showed no signed of mortality.

***Keywords**: Chlorophyll fluorescence, Photosynthesis, Rice, Submergence tolerance.*

Introduction

Complete submergence by flash flooding is a major production constraint in wide areas of south East Asia resulting in an increasing mortality of plants and low grain yield. The adverse effect of submergence is a result of various interrelated factors such as limited gas diffusion, reduced irradiance, decreases in adenylate energy charges, which causes several visible injuries and damages photosynthetic apparatus (Crawford and Breandle 1996; Drew 1997; Singh *et al.*, 2001). *In vivo* chlorophyll fluorescence has been used frequently in the past as a convenience and non-intrusive method to determine the tolerance of different species to different environmental conditions (Sayed 2003). Changes in chlorophyll fluorescence emission arising mainly from PS II provide information on almost all aspects of photosynthesis and as an efficient indicator of plant to different environmental stresses including submergence (Krause and Weis 1991; Sarkar *et al.*, 2004; Panda *et al.*, 2005). At present tolerant cultivars are identified by submerging them completely under water for a time period where mortality of the plants is the sole criterion to differentiate between tolerant and susceptible cultivar. The present work is an effort to characterize the submergence sensitivity in rice plants on the basis of the survival chance of the plant and of the changes in chlorophyll a fluorescence characteristics so that without mortality, tolerant and susceptible cultivar could be distinguished.

Hence, the goal of the present investigation is to know the effect of submergence on structural and functional alteration of PS II in rice and compare these responses between the genotypes to identify the possible selection criteria for submergence tolerance in rice.

Materials and Methods

Plant Material and Growth Conditions

The experiment was conducted on three *Indica* rice [*Oryza sativa* (L.)] cultivars namely, FR 13A (tolerant), IR 42 (susceptible) and Sabita (submergence avoiding type). All the cultivars were sown directly in earthen pots containing two kg of farm soil and farmyard manure in a 3:1 ratio. Each pot was supplied with 80mg urea, 192mg single super phosphate (P_2O_5) and 70mg murate of potash (K_2O). Ten days after germination, seedlings were thinned and five plants per pot were maintained. Twenty one-day-old seedlings were completely submerged for 10 days in a concrete tank under 110cm depth of water and one more set kept out side under normal condition served as control.

The pH of the floodwater was alkaline in nature (7.2-7.5), whereas the temperature varied between 25.2 to 31.4°C during the study period. Light intensity at 60cm depth of water was only 39.2–44.9 per cent compared to the air and oxygen concentrations were 2.32–3.45 mg L^{-1} at 0630 h and 5.15–7.48 mg L^{-1} at 1730 h.

Chlorophyll Estimation

For chlorophyll estimation, 100mg of finely chopped fresh leaves were placed in a 25ml capped measuring tube containing 25ml of 80 per cent acetone, and kept inside a refrigerator (4°C) for 28h (Sarkar 1998). The chlorophyll was measured spectrophotometrically following (Porra 2002).

Chlorophyll Fluorescence

Chlorophyll fluorescence was measured using a Plant Efficiency Analyzer, Handy PEA (Hansatech Instruments Ltd., Norfolk, UK). The chlorophyll fluorescence transients were induced by a red light of 3000 μE m^{-2} s^{-1}, and recorded from 10 μs up to 1s. All the measurements were done on fully dark-adapted attached leaves. From the first O-J-I-P transients, several bio-energetic parameters were derived according to the equations of JIP-test using BIOLYSER program (R. M. Rodriguez, Bioenergetic Laboratory, University of Geneva; available at http://www.unige.ch/sciences/biologie/bioen). For each treatment, the Chlorophyll a fluorescence transients of 12 individual leaves were measured. Different chlorophyll fluorescence parameters were analyzed by the so-called JIP- test (Strasser and Stresser 1995; Stresser *et al.,* 1995). The JIP- test represents translation of original data to biophysical parameters that quantify the energy flow through PS II. Different fluorescence parameters are shown in Table 3.1.

Table 3.1: The JIP-test Formulae Using Data Extracted from the Fast Fluorescence O-J-I-P Transient

F_0	Minimal fluorescence - when all the reaction centers are open or in oxidized state
F_J	Fluorescence intensity at the J-step (at 2ms)
F_M	Maximal fluorescence - when all the reaction centers are closed or in reduced state
V_J	$(F_{2ms} - F_0)/(F_M - F_0)$
Area	Area between fluorescence curve F_0 and F_M
Fv/Fm= (1-Fo/Fm)	Potential photochemical activity of PS II
ETo/CS	(Fv/Fm).(1-Vj).Fo
RC/CS	(Fv/Fm).(Vj/Mo).Fo
Performance index (PI_{ABS})	(RC/ABS). [(Fv/Fm)/1-(Fv/Fm)]. [(1-Vj)/(1-(1-Vj)]

Statistical Analysis

Differences between various chlorophyll fluorescence parameters were compared by ANOVA using IRRISTAT (International Rice Research Institute, Philippines) software's least significant difference (LSD<0.05), as this is a good test for determining whether means were significantly different.

Result and Discussions

The extent of visible injuries caused due to submergence is commonly utilized as an indicator of their sensitivity to it. In this experiment three cultivars gave distinctly different response to submergence in terms of visible symptoms and survival. Tolerant cv. FR 13A shows survival of 100 per cent after 10d of submergence whereas it was below 15 per cent in IR 42 and Sabita but there was no mortality even in susceptible cultivar after 6d of submergence (data not shown).

Submergence decreased the total chlorophyll content in all the cultivars (Figure 3.1) but the magnitude of reduction was greater in submergence avoiding type cv. Sabita. Chlorophyll content was more or less equal between susceptible cv. IR 42 and tolerant cv. FR 13A up to 8d of submergence. Significant difference was notice between susceptible and tolerant cultivar after 10d of submergence. Reduction of chlorophyll during submergence is common in rice (Sarkar *et al.*, 1996; Das and Sarkar 2001) and degradation of chlorophyll shows the loss of basic function in plants. The present investigation showed that chlorophyll based screening technique was not so sensitive to differentiate cultivars at mild stress (Panda *et al.*, 2006).

The minimal fluorescence (Fo) did not changed significantly up to 8 d of submergence (Table 3.2). The differences were significantly lower only in susceptible and avoiding types of cultivars after 10 d of submergence. A decrease of Fo is a characteristics of PS II disorganization at the antennae pigment level and decrease efficiency of excitation trapping at the active center of PS II (Bjorkman

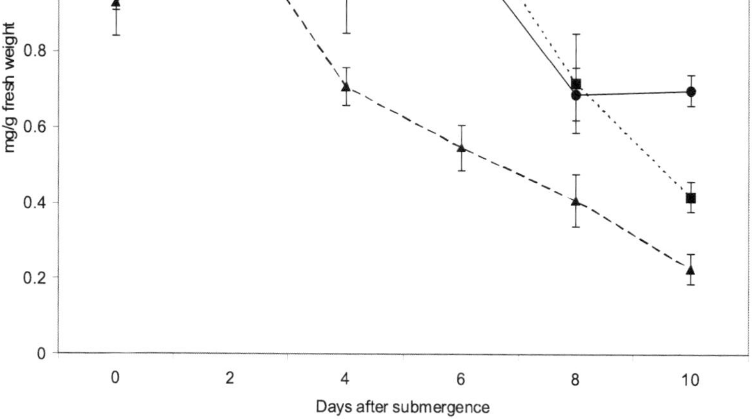

Figure 3.1: Changes in Chlorophyll Content During Submergence. Bar represents mean ÷ standard deviation

and Powel 1984; Panda et al., 2006). The maximum fluorescence (Fm) decreased significantly after 6 d of submergence in IR 42 and Sabita. In tolerant cv. FR 13A, the difference was significantly lower only after 10 d of submergence. The Fv/Fm ratio did not significantly decrease even after 10 d of submergence in tolerant cultivar whereas the differences were significantly lower after 6 d of submergence in susceptible and avoiding types of cultivars. The quenching of both the Fm and Fv/Fm or preferential of quenching of Fv would indicate more extensive damage to reaction center so that charge recombination is prevented or may be decrease in the activity of water splitting enzymes (Lazar 2006). More decrease of Fm and Fv/Fm value in the IR42 and Sabita during submergence may be due to more structural damage of PS II (Waldoff *et al.*, 2002).

Table 3.2: Changes of Minimum (Fo), Maximum (Fm) Fluorescence and Potential Photochemical Activity of PS II (Fv/Fm) Due to Submergence. In a column mean followed by the common letters are not significantly different at the 5 per cent level by DMRT; the unit of each parameter is arbitrary. DAS: Days after submergence.

DAS	*Fo*			*Fm*			*Fv/Fm*		
	FR 13A	*IR 42*	*Sabita*	*FR 13A*	*IR 42*	*Sabita*	*FR 13A*	*IR 42*	*Sabita*
0	508^{a}	520^{a}.	500^{b}	2440^{ab}	2520^{a}	2310^{a}	0.791^{a}	0.792^{a}	0.788^{a}
2	493^{a}	510^{ab}	502^{b}	2608^{ab}	2732^{a}	2669^{a}	0.806^{a}	0.813^{a}	0.810^{a}
4	530^{a}	554^{a}	597^{a}	2799^{a}	2592^{a}	2544^{a}	0.810^{a}	0.786^{a}	0.763^{a}
6	514^{a}	489^{ab}	594^{a}	2606^{ab}	1978^{b}	1833^{b}	0.802^{a}	0.752^{ab}	0.661^{b}
8	494^{a}	537^{a}	535^{ab}	2644^{ab}	1776^{b}	1632^{b}	0.813^{a}	0.695^{b}	0.691^{b}
10	501^{a}	451^{b}	457^{b}	2245^{b}	1292^{b}	1038^{c}	0.775^{a}	0.620^{c}	0.524^{c}
LSD5%		74.4			378.6			0.0638	
CV		8.7			9.2			5.1	

The Area above the fluorescence curve between Fo and Fm is proportional to the pool size of electron acceptor on the reducing side of the PS II, which includes Q_A, Q_B, and PQ (Strasser *et al.*, 1995; Joliot and Joliot 2002) is highly sensitive to submergence (Table 3.3). Submergence significantly decreased the plastoquinone pool in all the genotypes from 2 d of submergence. The decrease was more in the susceptible and avoiding type cultivars than the tolerant cultivar. After 10 d of submergence the decrease of the value area was 56 per cent, 80 per cent and 76 per cent in FR 13A, IR 42 and Sabita respectively. The phenomenological energy fluxes like electron transport per cross section (ETo/CS) represents the re-oxidation of reduced Q_A via electron transport over a cross-section of active and inactive reaction center whereas RC/CS represents the active PS II reaction center per exited cross-section (Force *et al.*, 2003). The greater values of ETo/CS and RC/CS designate the better structural and functional integrity of PS II. Both ETo/CS and RC/CS decreased due to submergence. However, the magnitudes of decrease of these two parameters were greater in IR 42 and Sabita compared to the FR 13A. The decrease of electron transport may be due to the inactivation of acceptor side of the PS II and the decrease in number of active reaction centre in a PS II cross-section (RC/CS) may be due to the damage of donor side of PS II (Panda *et al.*, 2006). The initial stage of photosynthetic activity of a reaction center complex is regulated by three functional steps namely absorption of light energy, trapping of the excitation energy and the conversion of excitation energy to electron transport. Combining these three steps a multi parametric expression of over all performance index (PI) of PS II was calculated (Strasser and Tsimlli- Michael 2001). Submergence significantly decreased the PI in all the cultivars (Table 3). However, tolerant cultivar always maintained higher values of this parameter compared to susceptible and avoiding cultivars. Similar results were obtained in dark chilling effect on soybean (van Heerden *et al.*, 2003).

Based on *in vitro* studies, two mechanisms of photo inhibitory damage have been proposed (Ohad *et al.*, 2000). Acceptor side photo inhibition begins from the over-reduction of the primary quinine acceptor Q_A that occurs when the electron flow between Q_A^- and Q_B, the secondary quinine acceptor, is blocked. The other mechanism of photo inhibitory damage is the donor side inhibition (Govindachary *et al.*, 2004). It has been suggested that only one general mechanism of photo inhibition predominates *in vivo* (Anderson *et al.*, 1998), however under submergence both donor and acceptor sides got damaged and tolerant cultivars to some extent prevent the damage as evident by higher area, ETo/CS and RC/CS and performance index (Table 3.3).

Conclusion

Our result showed that Chlorophyll a fluorescence parameters provide a non-invasive and rapid method for investigation of stress

Table 3.3: Changes of Different Chlorophyll Fluorescence Parameters Like Area Above the Fluorescence Curve, Electron Transport per Cross-section (ETo/CS), Active Reaction Center per Cross Section (RC/CS) and Performance Index of Photosynthesis (PI) Due to Submergence. In a column mean followed by the common letters are not significantly different at the 5 per cent level by DMRT; the unit of each parameter is arbitrary. DAS: Days after submergence.

DAS	*PI*			*ETo/CS*			*RC/CS*			*Area*		
	FR13A	*IR42*	*Sabita*	*FR13A*	*IR42*	*Sabita*	*FR13A*	*IR42*	*Sabita*	*FR13A*	*IR42*	*Sabita*
0	0.993^{a}	0.933^{a}	0.900^{a}	189^{a}	193^{a}	182^{a}	142^{a}	140^{a}	141^{a}	35200^{a}	29366^{a}	24300^{a}
2	1.033^{a}	0.967^{a}	1.067^{a}	168^{a}	200^{a}	202^{a}	137^{a}	132^{a}	138^{a}	26000^{b}	26733^{a}	24000^{a}
4	0.920^{a}	0.540^{b}	0.510^{b}	187^{a}	140^{b}	142^{b}	145^{a}	131^{a}	136^{a}	24133^{b}	18000^{b}	15133^{b}
6	0.640^{b}	0.280^{c}	0.233^{c}	144^{b}	80^{c}	90^{c}	133^{ab}	107^{b}	117^{b}	18966^{c}	9533^{c}	11800^{ab}
8	0.647^{b}	0.230^{c}	0.153^{c}	111^{c}	78^{c}	79^{c}	131^{ab}	96^{b}	104^{b}	17300^{c}	8333^{c}	7766^{cd}
10	0.393^{c}	0.137^{c}	0.117^{c}	103^{c}	48^{c}	56^{d}	118^{b}	52^{c}	65^{c}	15600^{c}	5933^{c}	6433^{d}
LSD5%		0.210			20			17			4141	
CV		21.4			9.2			8.5			13.8	

effect on PS II. Since submergence induces the structural and functional alteration of PS II in rice and fluorescence parameters are differentially changes in the three contrasting genotypes during the progression of submergence treatment. Among them area, performance index, electron transport and reaction center per cross section could determine the susceptibility/tolerance of the cultivars even after 4 to 6 d of submergence.

Acknowledgements

Authors are grateful to National Agricultural Technology Project (NATP), Indian Council of Agricultural Research, New Delhi, India for financial help.

References

Anderson JM, Park YI and Chow WS (1998) Unifying model for the photoinactivation of photosystem II in vivo under steady-state photosynthesis. Photosynthesis Research 56: 1-13.

Bjorkman O, Powel SB (1984) Inhibition of photosynthetic reactions under water stress: interaction with light level. Planta 161: 490-504.

Crawford RMM, Braendle R (1996) Oxygen deprivation stress in a changing climate. J Experimental Botany. 47: 145-59.

Das KK, Sarkar RK (2001) Post flood changes on the status of chlorophyll, carbohydrate and nitrogen content and its association with submergence tolerance. Plant Archives 1: 15-19.

Drew MC (1997) Oxygen deficiency and root metabolism: injury and acclimation under hypoxia and anoxia. Annual Review Plant Physiology Plant Molecular Biology 48: 223-250.

Force L, Critchley C, van Rensen JJS (2003) New fluorescence parameters for monitoring photosynthesis in plants. Photosynthesis Research 90: 1-19.

Govindachary S, Bukhov NG, Joly D, Carpentier R (2004) Photosystem II inhibition by moderate light under low temperature in intact leaves of chilling sensitive and tolerant plants. Physiologia Plantarum 121: 322-333.

Joliot P, Joliot A (2002) Cyclic electron transport in plant leaf. Proceedings National Academic Sciences, USA 99: 10209-10214.

Krause GH, Weis E (1991) Chlorophyll fluorescence and photosynthesis the basics. Annual Review Plant Physiology Plant Molecular Biology 42: 313-349.

Lazar D (2006) The polyphasic chlorophyll a fluorescence rise measured under high intensity of exiting light. Functional Plant Biology 33: 9-30.

Ohad I, Sonoike K, Andersson B (2000) Photoinactivation of the two photosystems in oxygenic photosynthesis: mechanisms and regulations. In: Yunus M, Pathre U, Mohanty P, eds. robing Photosynthesis: Mechanisms, Regulation and Adaptation. London: Taylor and Francis pp-293-309.

Panda D, Rao DN, Sharma SG and Sarkar RK (2006) Submergence effect on rice genotypes during seedling stage: probing of submergence driven changes of photosystem 2 by chlorophyll a fluorescence induction O-J-I-P transients. Photosynthetica 44: 69-75.

Panda D, Sharma SG and Sarkar RK. (2005) Changes of chlorophyll fluorescence and CO_2 photo-assimilation in rice cultivars during submergence and subsequent re-aeration. Plant Science Research 27: 76-80.

Porra RJ (2002) The chequered history of the development and use of simultaneous equations for accurate determination of chlorophylls a and b. Photosynthesis Research 73: 149- 156.

Sarkar RK (1998) Saccharide content and growth parameters in relation with flooding tolerance in rice. Biologia Plantarum 40: 597-603.

Sarkar RK, Panda D, Rao DN, Sharma SG (2004) Chlorophyll fluorescence parameters as indicators of submergence tolerance in rice. International Rice Research Notes 29: 66-68.

Sarkar RK, De RN, Reddy JN, Ramakrishnayya G (1996) Studies on the submergence tolerance mechanism in relation to carbohydrate, chlorophyll and specific leaf weight in rice *Oryza sativa* L.). Journal of Plant Physiology 149: 623-625.

Sayed OH (2003) Chlorophyll fluorescence as a tool in cereal crop research. Photosynthetica 41: 321-330.

Singh HP, Singh BB, Ram PC (2001) Submergence tolerance of rainfed lowland rice; search for physiological marker traits. J Plant Physiology 158: 883-889.

Strasser BJ, Strasser RJ (1995) Measuring fast fluorescence transients to address environmental questions: the JIP-test. In 'Photosynthesis: from light to biosphere'. (Ed. P Mathis) pp. 977-980. (Kluwer Academic Publishers, Dordrecht).

Strasser RJ, Srivastava A, Govindjee (1995) Polyphasic chlorophyll a fluorescence transient in plants and cyanobacteria. Photochemistry and Photobiology 61: 32-42.

Strasser RJ, Tsimilli-Michael M (2001) Stress in plants, from daily rhythm to global changes, detected and quantified by the JIP-test. Chimie Nouvelle (SRC) 75: 3321-3326.

van Heerden PDR, Tsimilli-Michael M, Krüger GH, Strasser RJ (2003) Dark chilling effects on soybean genotypes during vegetative development; parallel studies of CO_2 assimilation, chlorophyll a fluorescence kinetics O-J-I-P and nitrogen fixation. Physiologia Plantarum 117: 476-491.

Waldhoff, D, Furch, B, Junk WJ (2002) Fluorescence parameters, chlorophyll concentration, and anatomical features as indicators for flood adaptation of an abundant tree species in central Amazonia: *Symmeria paniculata*. Environmental and Experimental Botany 48: 225-235.

Chapter 4

Relative Tolerance of Several Species of Cyanobacteria to Commercial Fertilizers

Amarendra Mishra [1] and Siba P. Adhikary [2*]

[1]*Banki Autonomous College, Banki, Orissa*

[2]*P.G.Department of Botany, Utkal University, Vani Vihar, Bhubaneswar – 751 004*

ABSTRACT

Relative tolerance of twelve different species of heterocystous cyanobacteria belonging to 6 different genera, *viz. Anabaena, Calothrix, Cylindrospermum, Nostoc, Scytonema and Westiellopsis* to various concentrations of commercial fertilizers, *viz.* urea (46 per cent N), ammonium sulphate (21 per cent N), NPK (15:15:15), DAP (18 per cent N, 46 per cent P) and SSP (16 per cent P) was studied. A wide variation in the tolerance level of the cyanobacterial species to different fertilizers was observed; variation was marked among the species of the same genus and between the genera to the type and the dose of the fertilizers. Growth of most of the organisms decreased in comparison to control in presence of recommended dose of urea and ammonium sulphate. Growth as well as chloro-phyll-*a* content of certain species was also adversely affected

* Corresponding Author: Telefax: 0674-2587389;
E-mail: adhikary2k@hotmail.com

when the cultures were supplemented with recommended dose of NPK, DAP and SSP. *Scytonema* sp. UU 2433 and both the species of *Calothrix* were sensitive to even the lower concentrations of all the fertilizers suggesting that presence of additional surface features like sheath layers outside the cell wall played least significant role in protecting cyanobacterial cells from the adverse effect of higher levels of fertilizer application.

Keywords: *Biofertilizer, Cyanobacteria, Fertilizer, Rice field, Tolerance.*

Introduction

Free living cyanobacteria are distributed through out the world and contribute to the nitrogen fertility of many ecosystems. The agronomic potential of these organisms in rice cultivation was first recognised in 1939 by De, who attributed the natural fertility of tropical rice fields to biological N_2-fixation by cyanobacteria. Since then, long-term fertility experiments and N_2-fixation measurements have confirmed the importance of cyanobacteria in maintaining a moderate but consistent rice production in fields receiving no N-fertilizer (Roger and Kulasooriya, 1980; Venkataraman, 1981; Roger and Watanabe, 1986). Bunt (1961) reported the absence of nitrogen fixing cyanobacteria in experimental plots of Australia which he considered due to application of higher doses of N-fertilizers in the rice fields. Several field experiments in India have also shown that application of N-fertilizers was inhibitory for nitrogen fixing cyanobacterial population build up in the rice fields (Mahapatra *et al.*, 1971). Through a systematic field trial at 22 locations in different regions of India, Pillai (1980) demonstrated the beneficial effect of BGA inoculation without N-fertilizer than when 25 kgN/ha was applied. This shows that commercial fertilizers used in routine application in rice fields have important ecological effects in addition to those usually intended.

Though heterocystous cyanobacteria have been found abundantly in the rice fields of India, there has been a little work on their responses to commercial fertilizers that are used for increasing crop productivity. Singh (1975) first reported the growth and heterocyst differentiation of 12 different cyanobacteria in presence of ammonium sulphate, urea, potash and superphosphate in laboratory culture. He found that higher concentrations of these fertilizers affected the growth but potash and superphosphate did not suppress heterocyst differentiation. Subsequently, Anand and Karuppusamy (1987) and Anand (1990) studied the physiological responses of three N_2-fixing cyanobacteria to commercial fertilizers. Kolte and Goyal (1989) and Suseela and Goyal (1995) studied the effect of graded concentrations of ammoniacal nitrogen on the growth and nitrogen fixation of several cyanobacterial species isolated from the rice fields of Vidarbha region of Maharastra and Andhra Pradesh respectively. All these reports showed that growth and nitrogen fixation of cyanobacteria respond differently to chemical nitrogen which also vary with species, source of the N-fertilizer and their concentrations used in the experiments.

In India, nitrogen fixing cyanobacteria are now used as biofertilizer for rice cultivation. Since commercial fertilizers has been shown to produce adverse effect on these organisms, it is essential to screen efficient cyanobacterial species from local rice fields which are tolerant to a broad range of fertilizers for their use as biofertilizer. There have been little research on the diversity response among cyanobacterial taxa to a wide variety of fertilizers. In order to find out whether there exist any inter- and intra-specific variation among the cyanobacteria isolated from rice fields to commercial fertilizers, studies on the relative tolerance of 12 species belonging to six different genera *viz. Anabaena, Calothrix, Cylindrospermum, Nostoc, Scytonema and Westiellopsis* to various concentrations of five different fertilizers *viz.* urea, ammonium sulphate, NPK, DAP and SSP, commonly used in the rice cultivation of the region was carried out.

Materials and Methods

Axenic culture of twelve different species of cyanobacteria *viz. Cylindrospermum* sp. UU 142, *Cylindrospermum* sp. UU 245, *Anabaena fertilissima* UU 146, *Anabaena variabilis* UU 147, *Nostoc sphaericum* UU 1413, *Nostoc commune* UU 1414, *Calothrix parietina* UU 1423, *Calothrix* sp. UU 2427, *Scytonema multiramosum* UU 1431, *Scytonema* sp. UU 2433, *Westiellopsis* sp. UU 1441 and *Westiellopsis* sp. UU 1443 isolated from rice fields were used in the experiments. Experiments were carried out in cotton stoppered test tubes containing 10 ml of medium and 1 ml of homogenized suspension of each organism (absorbance of the culture at 760 nm after inoculation = 0.02). Five different commercial fertilizers *viz.* urea (46 per cent N), ammonium sulphate (21 per cent N), NPK (15:15:15), DAP (18 per cent N and 46 per cent P) and SSP (16 per cent P) obtained from the local market and used. Fresh stock solutions of the fertilizers were prepared with double distilled water, filtered through Whatmann No.1 filter paper, sterilized separately and added aseptically to the culture medium in the experimental tubes to obtain the desired concentrations (urea: 5, 10, 25, 50, 75, 100, 200, 250, 500, 750, 1000 and 2000 μg/ml; ammonium sulphate: 10, 25, 50, 100, 250, 500, 750, 1000 and 2000 μg/ml; NPK: 10, 25, 50, 100, 250, 500, 750, 1000 and 2500 μg/ml; DAP: 25, 50, 100, 250, 500, 750, 1000, 2000, 2500 and 5000 μg/ml; SSP: 50, 100, 250, 500, 1000, 2000, 2500, 4000, 5000,7500, 10000 and 15000 μg/ml). After addition of various concentrations of fertilizers, pH of the medium was adjusted to 7.5 aseptically using sterile NaOH solution. Cultures were incubated up to 15 days at 25±1°C under continuous light at an intensity of 7.5 W/m^2 provided from day light fluorescent tubes.

For growth measurement, 3 tubes were removed at 2 day intervals, the contents were homogenized and absorbance of the culture suspension was read at 760 nm. Chlorophyll-*a* content of the cultures on the 15th day of their incubation was determined following Mackinney (1941). Median lethal dose (EC_{50}) of the fertilizers after 15 days of growth period of each organism was inversely estimated from the regression lines of percentage inhibition of growth on log dose of the fertilizers. Mean value of triplicate determinations ± S.D. has been presented.

Results

Growth of 12 species of rice field cyanobacteria belonging to 6 different genera in presence of various concentrations of five different commercial fertilizers was determined. Wide variation in the EC_{50} and lethal concentration of the commercial fertilizers to the cyanobacterial species was observed (Figures 4.1–4.5). Lethal dose varied from 100-2000, 100-2000, 1000-2500, 1000-5000, and 2500-15000 μg/ml for urea, ammonium sulphate, NPK, DAP and SSP respectively. There was no indication that tolerance capacity of a particular species or both the species under the same genus was equal to all the fertilizers. In general, *Scytonema* UU 2433 and both the species of *Calothrix*, UU 2427 and UU 1423 were sensitive to all the fertilizers; *A. fertilissima* UU 146 to ammonium sulphate and urea; *A. variabilis* UU 147 to urea, ammonium sulphate, NPK and

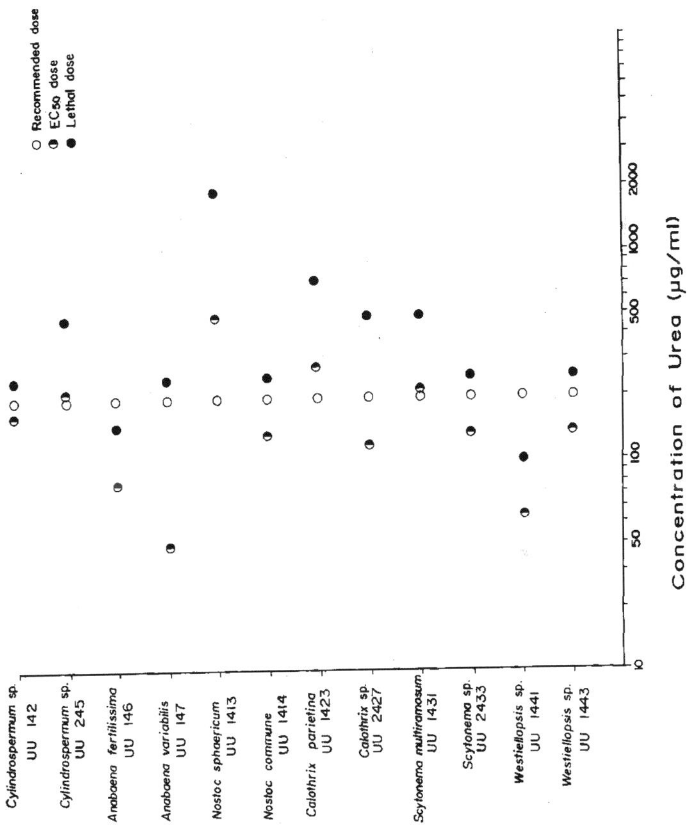

Figure 4.1: Recommended, EC_{50} and Lethal Dose (□ g/ml) of Urea on the Growth of Twelve Different Species of Rice Field Cyanobacteria

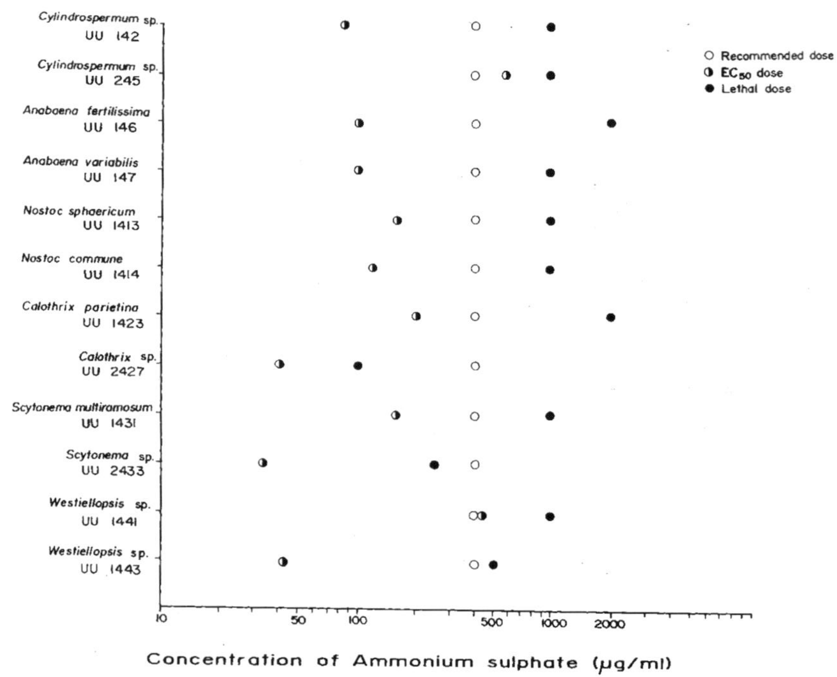

Figure 4.2: Recommended, EC_{50} and Lethal Dose (□ g/ml) of Ammonium Sulphate on the Growth of Twelve Different Species of Rice Field Cyanobacteria

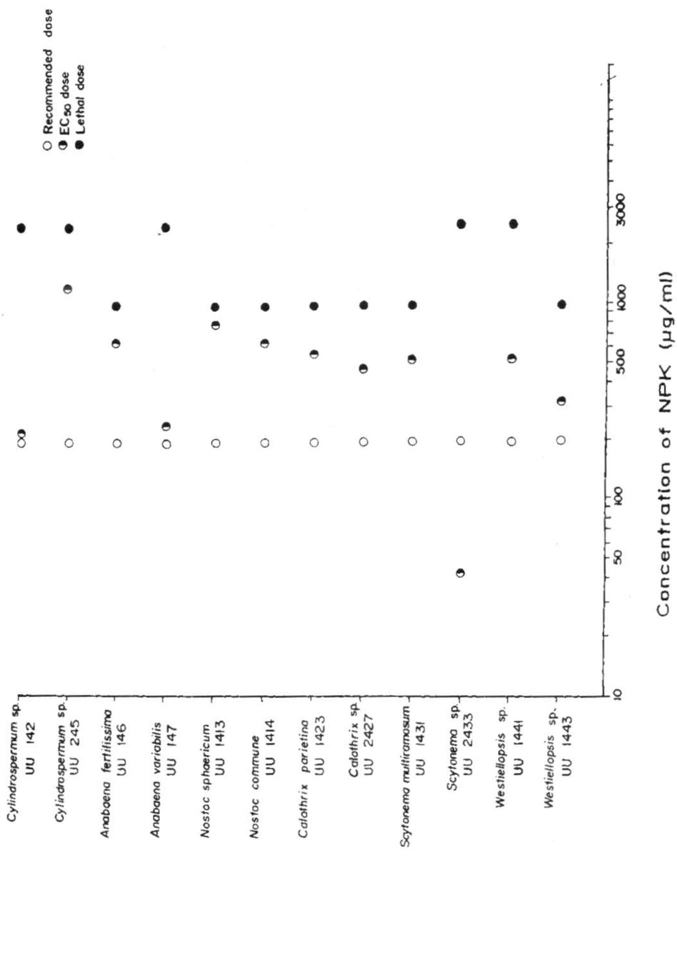

Figure 4.3: Recommended, EC_{50} and Lethal Dose (□ g/ml) of NPK on the Growth of Twelve Different Species of Rice Field Cyanobacteria

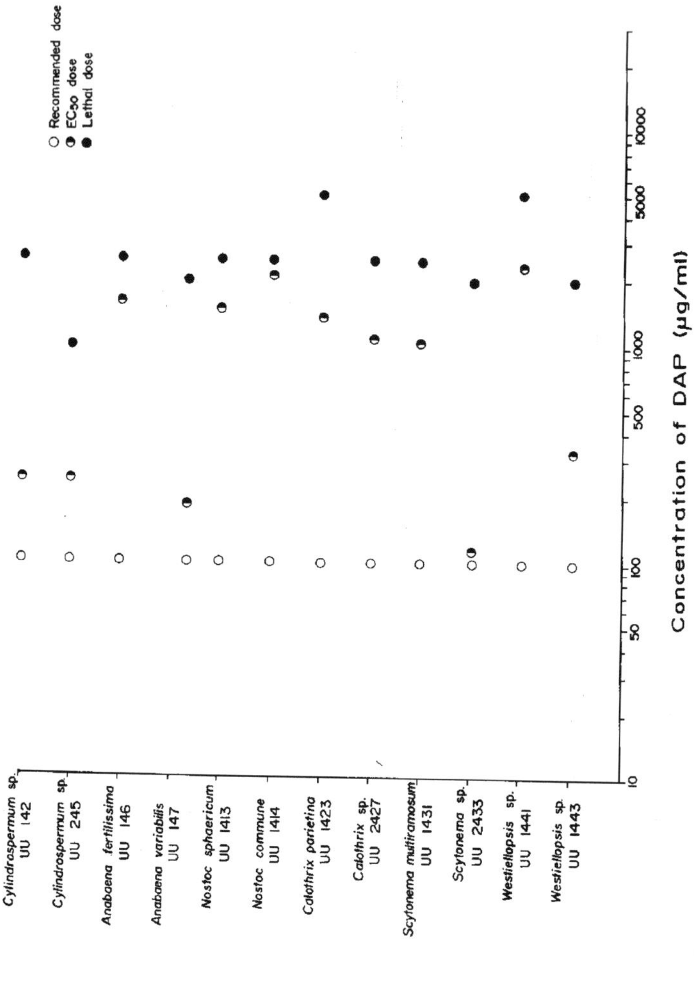

Figure 4.4: Recommended, EC_{50} and Lethal Dose (□ g/ml) of DAP on the Growth of Twelve Different Species of Rice Field Cyanobacteria

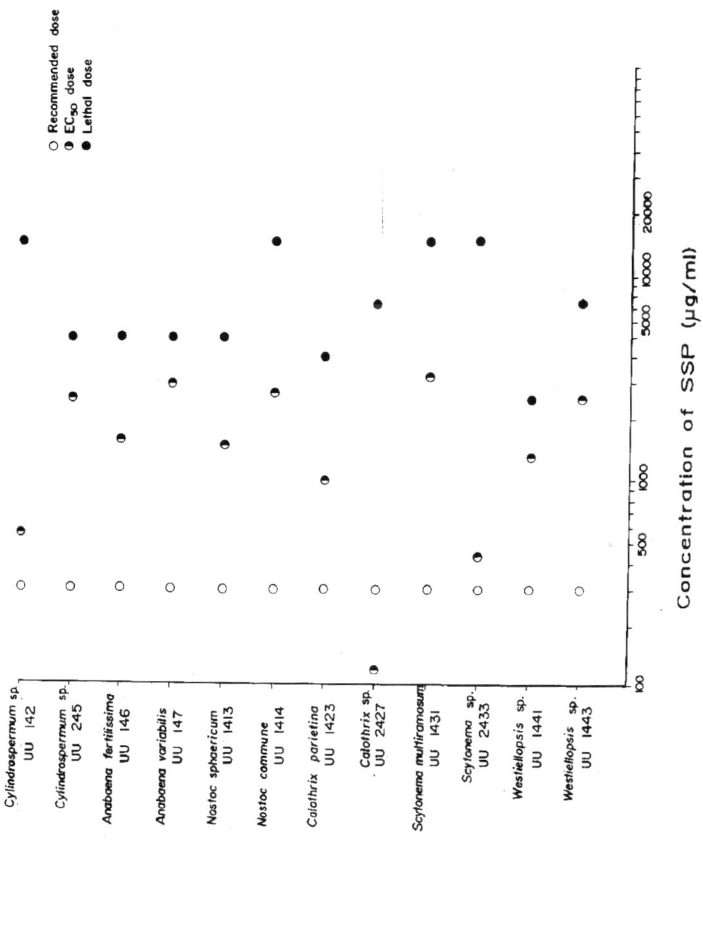

Figure 4.5: Recommended, EC_{50} and Lethal Dose (□ g/ml) of SSP on the Growth of Twelve Different Species of Rice Field Cyanobacteria

DAP; *Cylindrospermum* UU 142 to ammonium sulphate, NPK, DAP and SSP; *Nostoc commune* UU 1414 to ammonium sulphate; *Scytonema multiramosum* UU 1431 to ammonium sulphate, NPK and DAP and *Westiellopsis* sp. UU 1441 to urea, NPK and SSP was observed. However, *Cylindrospermum* sp. UU 245 grew well in presence of comparatively higher concentrations of urea, ammonium sulphate, NPK and SSP though the organism was extremely sensitive to DAP. Similarly, though *N. sphaericum* UU 1413 tolerated maximum concentration of urea and SSP, it could not do so to the same extent when grown with ammonium sulphate, NPK and DAP. It was observed that a particular species of the same genus though was comparatively more tolerant to a specific fertilizer, the other did not show better growth in presence of the same fertilizer (Table 4.1). Chlorophyll-*a* content of the organisms after cultivating in the various concentrations of the fertilizer containing media for 15 days was also estimated. Per cent increase or decrease in the chlorophyll-*a* content in presence of fertilizers over the control value of the respective organism was plotted (Figures 4.6-4.10). Analysis of these results showed almost a similar increase or decrease of chlorophyll-*a* content over the control in presence of a specific fertilizer as that of the respective growth values. Chlorophyll-*a* content of *Cylindrospermum* sp. UU 245 increased to about 120, 50, 325 and 200 per cent over control in presence of 50 µg/ml of urea, ammonium sulphate and 250 µg/ml of NPK and SSP respectively but decreased to over 50 per cent even when 100 µg/ml of DAP was supplemented to the culture. Chlorophyll-*a* of *N. sphaericum* UU 1413 increased to 150 per cent over control in presence of 150 µg/ml urea; other nitrogen sources except DAP did not support its chlorophyll-*a* synthesis comparable to control. Similarly, chlorophyll-*a* content of *Westiellopsis* sp. UU 1441 was increased by 72 per cent and 67 per cent in presence of 50 µg/ml and 250 µg/ml of ammonium sulphate and DAP respectively; all other nitrogenous fertilizers decreased its content. In general, chlorophyll-*a* of most of the species were adversely affected in presence of ammonium sulphate followed by NPK and urea. DAP containing 46 per cent P in addition to 18 per cent N and SSP with 16 per cent P did increase the chlorophyll-*a* content of most of the species at low concentrations except of *Scytonema* sp. UU 2433 and *Calothrix* sp. UU 2427. However, DAP containing nitrogen in addition to phosphate adversely affected the chlorophyll-*a* of cyanobacteria more than that of SSP fertilizer treatment.

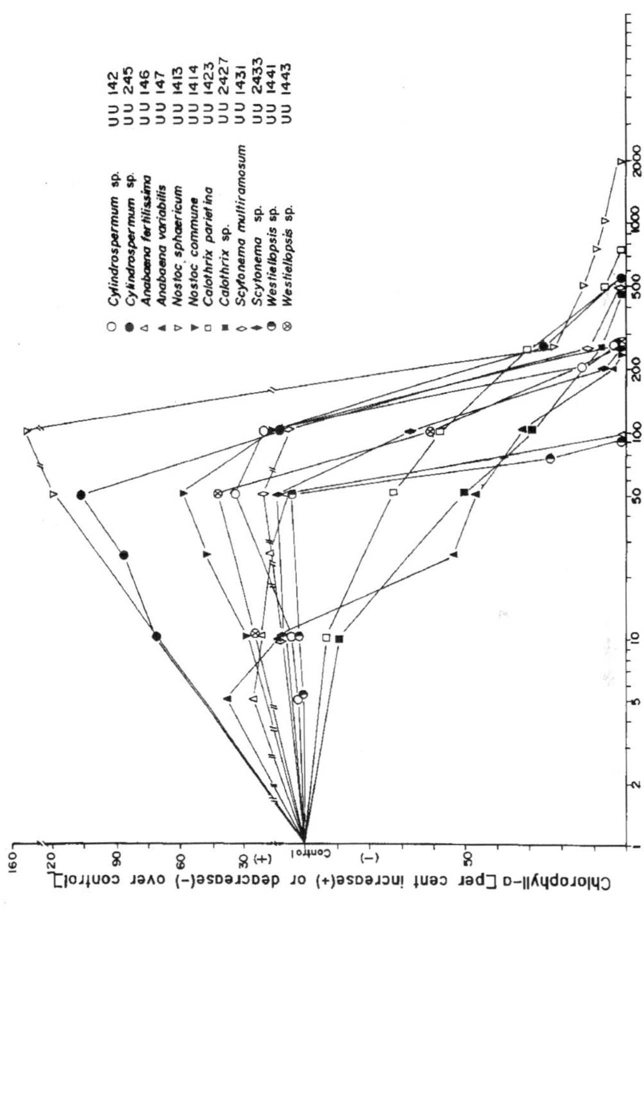

Figure 4.6: Chlorophyll-*a* Content [% Increase (+) or Decrease (-) Over Control Value] of Twelve Different Species of Cyanobacteria in Presence of Different Concentrations of Urea

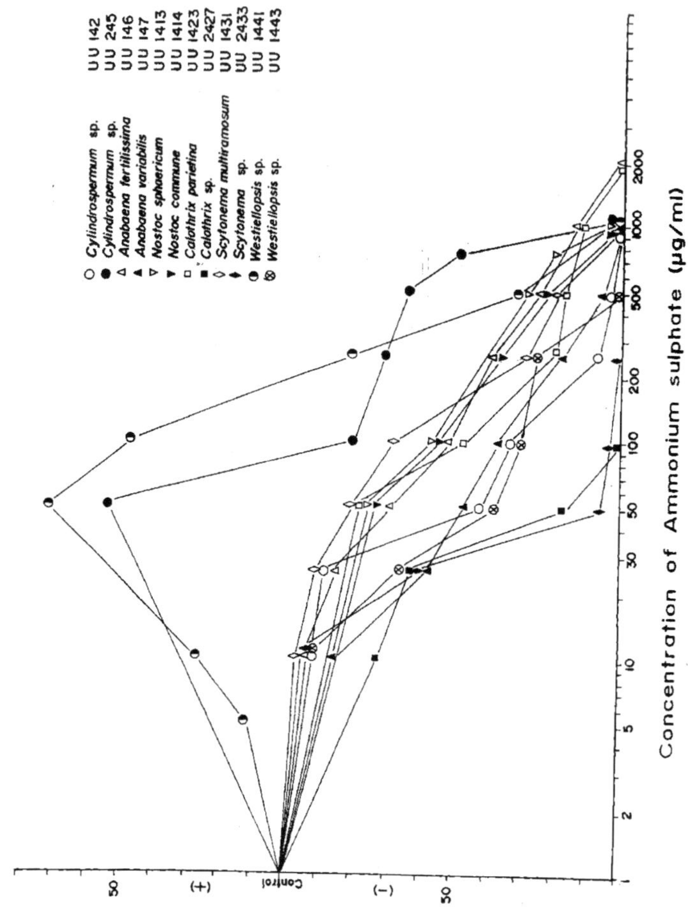

Figure 4.7: Chlorophyll-*a* Content [% Increase (+) or Decrease (-) Over Control Value] of Twelve Different Species of Cyanobacteria in Presence of Different Concentrations of Ammonium Sulphate

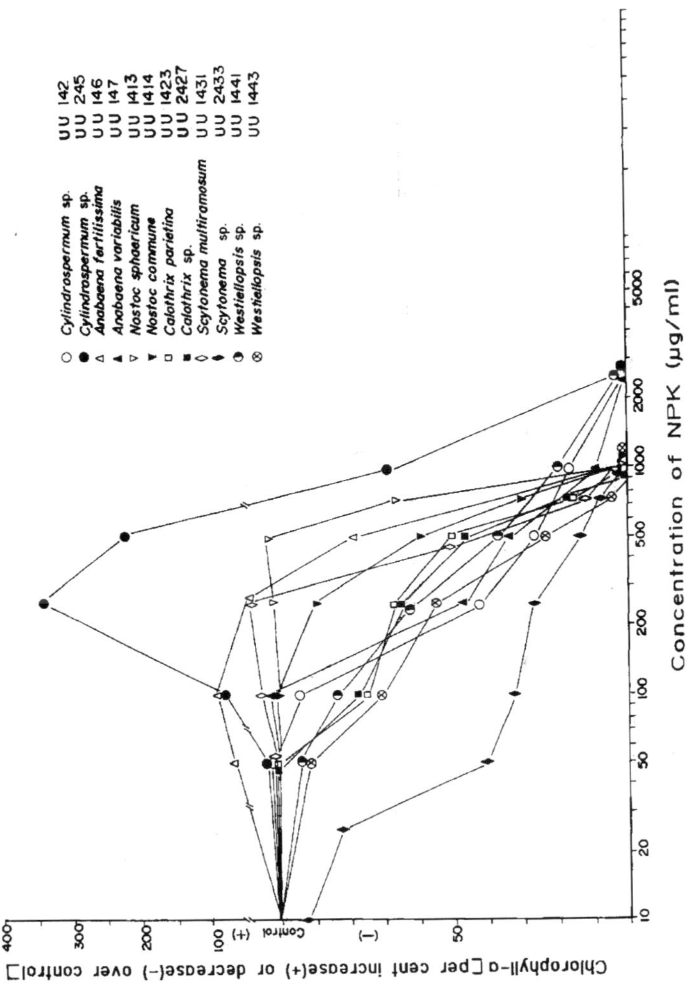

Figure 4.8: Chlorophyll-*a* Content [% Increase (+) or Decrease (-) Over Control Value] of Twelve Different Species of Cyanobacteria in Presence of Different Concentrations of NPK

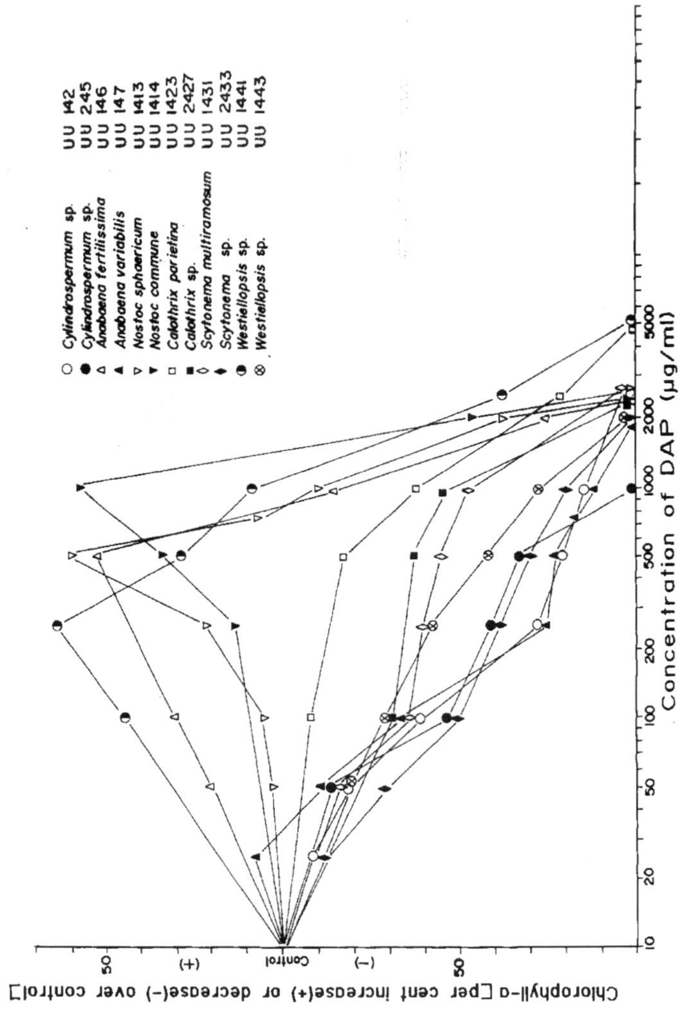

Figure 4.9: Chlorophyll-*a* Content [% Increase (+) or Decrease (-) Over Control Value] of Twelve Different Species of Cyanobacteria in Presence of Different Concentrations of DAP

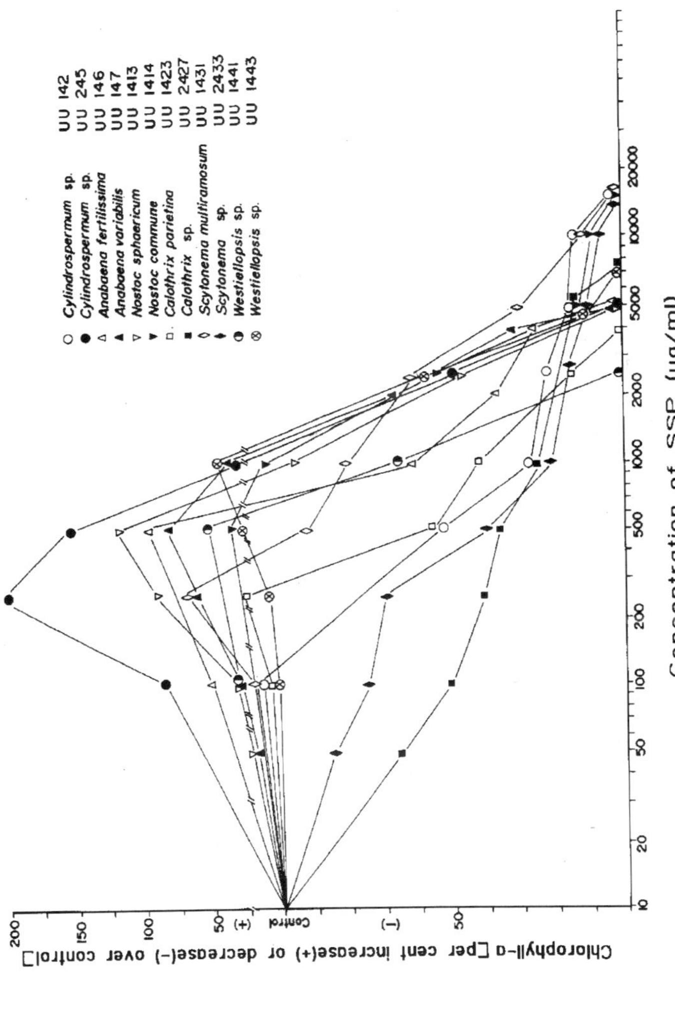

Figure 4.10: Chlorophyll-*a* Content [% Increase (+) or Decrease (-) Over Control Value] of Twelve Different Species of Cyanobacteria in Presence of Different Concentrations of SSP

Table 4.1: Tolerance of Various Rice Field Cyanobacteria to Commercial Fertilizers

Fertilizer	*Order of Tolerance*
Urea	*Nostoc* 1413> *Calothrix* 1423> *Scytonema* 1431> *Cylindrospermum* 245> *Calothrix* 2427> *Cylindrospermum* 142> *Scytonema* 2433 = *Westiellopsis* 1443> *Nostoc* 1414> *Anabaena* 147> *Westiellopsis* 1441> *Anabaena* 146.
Ammonium sulphate	*Calothrix* 1423> *Anabaena* 146> *Cylindrospermum* 245> *Westiellopsis* 1441> *Scytonema* 1431 = *Nostoc* 1413> *Nostoc* 1414> *Anabaena* 147> *Cylindrospermum* 142> *Westiellopsis* 1443> *Scytonema* 2433> *Calothrix* 2427.
NPK (15:15:15)	*Cylindrospermum* 245> *Westiellopsis* 1441> *Anabaena* 147> *Cylindrospermum* 142> *Scytonema* 2433> *Nostoc* 1413> *Nostoc* 1414 = *Anabaena* 146> *Calothrix* 1423> *Calothrix* 2427> *Scytonema* 1431> *Westiellopsis* 1443.
DAP (Di-amino phosphate)	*Westiellopsis* 1441> *Calothrix* 1423> *Nostoc* 1414> *Anabaena* 146> *Nostoc* 1413> *Calothrix* 2427> *Scytonema* 1431> *Cylindrospermum* 142> *Westiellopsis* 1443> *Anabaena* 147> *Scytonema* 2433> *Cylindrospermum* 245.
SSP (Single super phosphate)	*Scytonema* 1431> *Nostoc* 1414> *Cylindrospermum* 142> *Scytonema* 2433> *Cylindrospermum* 245> *Westiellopsis* 1443> *Calothrix* 2427> *Anabaena* 147> *Anabaena* 146> *Nostoc* 1413> *Calothrix* 1423> *Westiellopsis* 1441.

Discussion

Nitrogen fixing cyanobacteria have received much attention for their role in supplying nitrogen to rice crop (Singh, 1961; Venkataraman, 1981; Singh, 1985). In the recent years consumption of fertilizers has increased considerably with the development of intensive agriculture. Since diazotrophic cyanobacteria of waterlogged rice fields directly come in contact with various fertilizers soon after their application, they are bound to interact with these agrochemicals in their natural habitat (Bisoyi and Singh, 1988). Thus it was considered important to evaluate the effect of these fertilizers in the laboratory on pure cultures of cyanobacteria isolated from these fields.

It is already known that combined nitrogen inhibits nitrogen fixation but the degree and type of inhibition depends on the level supplied (Roger and Kulasooriya, 1980). Stewart *et al.* (1975) studied the effect of nitrogen fertilizers on *Anabaena cylindrica* and concluded that addition of such fertilizers to the natural population of cyanobacteria will affect the nitrogen fixation, and the depressive effect of ammonium-nitrogen is likely to be more than the nitrate-nitrogen, Singh (1975) studied the effects of N, P, K fertilizers on the growth and nitrogen fixation of 12 cyanobacterial species and found that (1) higher doses of fertilizers affected growth although the response varied from species to species, (2) addition of 10 mg/ml of ammonium sulphate and urea did not affect growth and heterocyst differentiation; 100 mg/ml and higher concentrations of ammonium sulphate showed reduction in growth and suppression of heterocyst frequency, and (3) the organisms tolerated higher concentration of potash than superphosphate although sensitivity varied from species to species. Pattnaik and Singh (1978) observed that calcium nitrate up to 3000 mg/ml did not inhibit growth of *Gloeotrichia* sp. They concluded that addition of NO^{-}_{3} fertilizers could be more beneficial in the rice fields as far as nitrogen fixing activity is concerned. However, use of nitrate fertilizers are not recommended for rice due to rapid rate of denitrification as well as leaching losses in the waterlogged conditions.

The work so far available showed that N_2-fixing cyanobacteria preferred lower concentration of nitrogenous fertilizers in milieu for good growth (Singh, 1975; Kolte and Goyal, 1989; Anand, 1990). However, there are also species which unable to show any improvement in their growth or metabolic activities in presence of any of the N-fertilizers tried (Anand, 1990). External source of nitrogen has been rather optional in many specieses which grow better in nitrogen fixing condition (Fogg *et al.*, 1973). Certain cyanobacterial strains possess the capacity to carry out nitrogen fixation unabated even in the presence of a nitrogen source. Though in the presence of external source of nitrogen the process of nitrogen fixation may be redundant, these strains exhibit a unique behaviour. It was in *Anabaena* sp. CA, Bottomley *et al.* (1979) found nitrogenase activity in the presence of high ammonium. Thus studies of the interaction between prospective biofertilizer containing efficient nitrogen fixing species and the existing chemical fertilizers is of great significance as the former which are known to supplement the latter now may ultimately substitute them in the rice fields at a future date.

Results obtained in the present experiment showed that there exist a remarkable variation in the growth behaviour of rice field cyanobacteria perceptible even at intraspecific level such as that screening of natural population can lead to the identification of cyanobacterial strains making ideal inoculants compatible with high level of fertilizers. Earlier reports on the tolerance of several nitrogen fixing cyanobacteria of rice fields to the insecticide Furadan showed that those forms which possess a well defined sheath layer outside their cell wall, such as those belong to *Calothrix and Scytonema*, were comparatively more tolerant than the sheathless forms (Rath and Adhikary, 1994). However, results of the present investigation showed that presence of additional surface features around the trichome of a cyanobacterium play least role to protect the adverse effect of fertilizers on any specific organism.

The recommended dose of urea, ammonium sulphate, NPK, DAP and SSP for rice crop is 120, 240, 120, 60 and 180 Kg/ha respectively. At these doses, the concentration of these fertilizers in the rice field water will be as follows: urea 200 µg/ml, ammonium sulphate 400 µg/ml, NPK 200 µg/ml, DAP 100 µg/ml and SSP 300 µg/ml which of course vary depending on the water level during the paddy cultivation cycle. The result of the present investigation showed that growth of most of the organisms decreased in comparison to control even in presence of the recommended dose of ammonium sulphate and urea. Growth of *Calothrix* sp. UU 2427, *Scytonema* sp. 2433 and of *Anabaena fertilissima* UU 146 and *Westiellopsis* sp. UU 1441 was completely stopped and the cells became bleached in presence of recommended dose of ammonium sulphate and urea respectively. Growth of almost all the organisms was decreased by over 50 per cent when the same dose of the fertilizers was supplemented to the cultures. Recommended dose of NPK and SSP was toxic to *Scytonema* UU 2433 and *Calothrix* sp. UU 2427 respectively. Growth as well as chlorophyll-*a* content of certain species was also adversely affected when the cultures were supplemented with recommended dose of NPK, DAP and SSP. From these results it is apparent that application of nitrogenous fertilizers even at the recommended dose reduced the growth of rice field cyanobacteria there by suppressing the nitrogen gain by natural population of these organisms in the rice field environment.

Acknowledgements

We thank the Head of the Department of Botany for laboratory facilities. One of us (AM) thank the Department of Higher Education,

Government of Orissa and the Principal, Athagarh College for grant of study leave to carry out this work at Utkal University.

References

Anand N (1990) Physiological responses of nitrogen fixing blue-green algae (cyanobacteria) to commercial fertilizers. In: Proc. Natl. Symp. Cyanobacterial Nitrogen Fixation. Ed. Kaushik B D, Associated Publishing Co. New Delhi, pp. 83-90.

Anand N and Karuppusamy A (1987) Growth of nitrogen fixing and non-nitrogen fixing blue-green algae in presence of some common fertilizers. Phykos 26: 22-26.

Bisoyi RN and Singh PK (1988) Effect of phosphorus fertilization on blue-green algal inoculum production and nitrogen yield under field condition. Biol. Fertil. Soils 5: 338-343.

Bottomley PJ, Grillo JF, Van Baalen C and Tabita FR (1979) Synthesis of nitrogenase and heterocysts by *Anabaena* sp. CA. in the presence of high levels of ammonia. J. Bact. 140: 938-943.

Bunt JS (1961) Nitrogen-fixing blue-green algae in Australian rice soils. Nature 192: 479-480.

De PK (1939) The role of blue-green algae in nitrogen fixation in rice fields. Proc. Roy. Soc. B. London 127: 121-134.

Fogg GE, Stewart WDP., Fay P and Walsby AE (1973) The Blue-Green Algae. Academic Press, New York and London, 459 p.

Kolte SO and Goyal SK (1989) Effect of ammonium nitrogen on growth and uptake and fixation of nitrogen by blue-green algae. Acta Bot. Ind. 17: 172-176.

Mackinney G (1941) Absorption of light by chlorophyll solutions. J. Biol. Chem. 140: 315-322.

Mahapatra IC., Mudholkar NJ and Pattnaik H (1971) Effect of N, P and K fertilizers on the prevalence of algae under field conditions. I.J. Agron. 14: 19-22.

Pattnaik U and Singh PK (1978) Effect of nitrate nitrogen on the growth, heterocyst differentiation and nitrogen fixation in rice field blue-green alga *Gloeotrichia* sp. Arch. Hydrobiol./Suppl. 51: 318-327.

Pillai KG (1980) Biofertilizers in rice cultures. Problems and prospects for large scale adoption. All India Coordinated Rice Improvement Project. All-India Rice Research Workshop, Bangalore, Publication no. 196.

Rath B and Adhikary SP (1994) Relative tolerance of several nitrogen-fixing cyanobacteria to commercial grade Furadan (Carbofuran, 3 per cent). I.J. Expt. Biol. 32: 213-215.

Roger PA, and Watanabe I (1986) Technologies for utilizing biological nitrogen fixation in lowland rice: potentialities, current usage, and limiting factors. Fert. Res. 9: 39-77.

Singh RN (1961) The role of Blue-green Algae in Nitrogen Economy of Indian Agriculture. Indian Council of Agricultural Research, New Delhi, 175 p.

Singh PK (1975) Fertilizers tolerance of blue-green algae and their effect on heterocyst differentiation. Phykos 14: 81-88.

Singh PK (1985) Nitrogen fixation by the blue-green algae in paddy fields. In: Rice Research in India. Eds. Jaiswal PL, Wadhwani AM, Singh R, Chhabra NN and Chhabra N., ICAR, New Delhi, pp. 344-362.

Stewart WDP, Haystead A and Dharmawardene MWN (1975) Nitrogen assimilation and metabolism in blue-green algae. In: Nitrogen fixation by free-living microorganisms. Ed. Stewart WDP, Cambridge Univ. Press, London, pp. 129-58.

Suseela MR and Goyal SK (1995) Effect of ammonium nitrogen on growth and nitrogen fixation by cyanobacteria. Phykos, 34: 123-130.

Venkataraman GS (1981) Blue-green algae for rice production: A manual for its promotion. FAO/UNDP Project document no. 2 (RAS/75/004), FAO, Rome, 102p.

Chapter 5

Effect of Phosphatic Fertilizers and Pasture Legumes on Forage Production of Trisankar (Pennisetum hybrid) Pasture

Keshab K. Mohapatra [1], *M.K. Das* [2], *Vinod Shankar* [3]

[1]*Assistant Fodder Development Officer, Chief District Veterinary Office, Sambalpur*

[2]*P.G. Department of Botany, G.M.College (Autonomous), Sambalpur*

[3]*Indian Grassland and forage Research Institute, Jhansi*

ABSTRACT

The grassland areas are decreasing due to the rapid urbanization and crop husbandry. The carrying capacity of the pasture is also declining due to heavy grazing pressure and other biotic interference. There is no hope that the farmers will bring more cultivable area under fodder productivity. The only answer is grasslands which if improved and managed properly could substantially provide the required herbage. The low protein value and poor herbage yield of tropical grasses can be raised by the application of phosphatic fertilizer and legumes.

In the present experiment, an attempt has been made to know the effect of levels of phosphatic fertilizer with and without pasture legumes on the forage production of the Trisankar pasture. Though the fresh forage yield was less in the grass + legume + phosphorous pasture than the grass + phosphorous pasture, the dry forage was higher in grass + legume + phosphorous pasture than the grass + phosphorous pasture. This can possibly be ascribed to better nodulation of legumes due to the application of phosphatic fertilizer which ultimately resulted in the higher dry matter productions.

Keywords: *Forage, Pasture, Phosphatic fertilizers, Pasture legumes.*

Introduction

Since time immemorial, animal husbandry continues to be the most important profession of Indian farmers. In the past, grasslands and other grazing areas were very productive and could sustain the entire livestock population. Gradually the pressure of the human population increased and this resulted to the enhanced demand of livestock products like milk, meat, wool etc. The urbanization and crop husbandry became the major reasons for the decline in grassland area, which led to decline in the carrying capacity of grasslands and subsequently what we see today is the vast stretches of degraded rangelands. The present condition of grasslands is extremely poor in terms of the herbage availability and it is being further reduced due to weed infestation, erosion of top soil etc.

Singh *et al.* (1984) has calculated that the country is reaching only 46.6 per cent of the total requirement, although there are certain states in the country like Punjab, Haryana and Utter Pradesh, where fodder is cultivated. Only 4.4 per cent of the country's total cropping area is put under fodder crops. Cereals and other cash crops still remain first priority of the farmers. For livestock, the farmers still depend on grasslands and other grazing areas. This situation has been continuing since long and the grazing lands may further deteriorate unless corrective measures are taken immediately. The area of the grasslands can not be increased. The unproductive livestock can not be eradicated because of socio-economic and religious believes. The only answer is grasslands which if improved and managed properly could substantially provide the required herbage. Keeping this in view the present experiment was conducted to observe the effect of different levels of phosphorous in the established pasture of Trisankar (*Pennisetum* hybrid) with inter-cropping of two legumes *viz.* Caribbian Stylo (*Stylosanthes hamata* and Siratro (*Macroptilium atropurpureum*).

Forage plants have been extensively studied and researched throughout the world. However, in India, this is comparatively a recent subject and the earliest paper written about the cultivation of fodder seems to be that of Robbertson Brown (1914) in which the possibilities of Ryegrass and Clover cultivation in India have been discussed. Burns (1915) stressed on the need for improvement of natural grasslands of India. Bharucha (1958) evaluated the problems of grassland improvement in India. Ever since various studies on different aspects of forage plants and grasslands have been continuing by various workers (Shankarnarayan *et. al.*, 1975; Velayudhan *et. al.*, 1977; Rai *et. al.*, 1980; Prasad and Mukherjee, 1980, Rai and Patil, 1985, Dwivedi *et. al.*, 1985; Niranjan *et. al.*, 1990; Kanodia and Patil, 1991).

Materials and Methods

Experimental Site

The field trial was conducted at Central Research Farm of IGFRI (Indian Grassland and Fodder Research Institute), Jhansi (MP). The Research farm is situated at 78° 35'E, 25° 26' N and about 275 MMSL. The experiment was conducted under rainfed condition. The soil of the experimental field was sandy loam, low in phosphorous (4.7 kg P_2O_5/ha), medium in available N (252 kg/ha) and available potash (225 kg/ha) and pH is 6.8.

Experimental Materials

The grass selected, Trisankar (*Pennisetum* hybrid) is a cross among *P. typhoides*, *P. purpureum* and *P. squamulatum* by U.S.A. plant breeder Dr. Hana. It was introduced at IGFRI during 1988. It is a tall, tussocky, perennial grass producing profuse tillers and apomyctic seeds. It is an ideal choice for rainfed conditions for the revegetation of degraded rangelands.

The Caribbian stylo (*Stylosanthes hamata*), originated on the islands of West Indies and the adjacent coastal regions of the North and South America continent, was collected by Mr. Tom Atkinson of N.S.W., Department of Agriculture, Venezuela. It is an erect, smooth stemmed herbaceous plant. It grows on a wide variety of soil type and is more drought tolerant. It performs well in the annual rainfall range of 500-1270mm.

Siratro (*Macroptilium atropurpureum*) was bred by Dr. E.M. Hutton, Division of Tropical Pasture, C.S.I.R.O., Australia, from two parent lines collected in Mexico. The seeds of siratro has been proved to be one of the most productive and adoptable tropical legumes throughout the world. It is a deep rooted perennial with trailing, slightly hairy stems with root readily at the nodes. Leaves are trifoliate, dark green and slightly hairy on the upper and silvery and very hairy on the lower surface. It is very much suitable for tropical to sub-tropical climate with an annual rainfall of 760 to 1780 mm.

Methodology

For the present study the 4 plots of every replication were treated with different levels of phosphorus. Next 4 plots of the every replication was intercropped by stylo by uprooting the middle line of every three lines of the grasses and treated with different doses of phosphorus. Similarly the last 4 plots of every replication was intercropped by siratro and treated with phosphorus (Figure 5.1)

Observations

Observations on the following parameters were taken:

Plant Population

Plant population of Trisankar Pennisetum, Stylo and Siratro were counted from 1x1 m quadrant from each treatment (grass and legume separately) and averaged at 30 days and 60 days of sowing of legumes.

Plant Height

The height of the plant of the grass and legumes measured by using scale from 5 randomly selected plants in each treatment and averaged at 30 days and 60 days of sowing of legumes.

Number of Branches/Tillers

The number of branches of legumes and tillers of grass were counted from 5 randomly selected plants in each treatment and averaged at 30 days and 60 days of sowing of the legumes.

Tussock Diameter

The tussock diameter of the gram was measured from 5 randomly selected plants of averaged at 30 days and 60 days of showing of legumes.

Experimental details

Plot size	-	3 × 5 m
Replications	-	3
Levels of phosphorus	-	0, 30, 60, 90 kg P_2O_5/ha
Total No. of treatment	-	12
Design	-	Factorial R.B.D.
T_1	-	$G + P_0$
T_2	-	$G + P_{30}$
T_3		$G + P_{30}$
T_4	-	$G + P_{90}$
T_5	-	$G + L_a + P_0$
T_6	-	$G + L_a + P_{30}$
T_7	—	$G + L_a + P_{60}$
T_8	-	$G + L_a + P_{90}$
T_9	-	$G + L_b + P_0$
T_{10}	-	$G + L_b + P_{30}$
T_{11}	-	$G + L_b + P_{60}$
T_{12}	-	$G + L_b + P_{90}$
Notes	-	
L_a	-	Stylo (*Stylosanthes hamata*)
L_b	-	Siratro (*Macroptilium atropurpureum*)
G	-	Trisankar Pennisetum

R_1

T_1	T_2	T_3	T_4
T_6	T_7	T_8	T_5
T_{11}	T_{12}	T_9	T_{10}

R_2

T_8	T_5	T_6	T_7
T_9	T_{10}	T_{11}	T_{12}
T_2	T_3	T_4	T_1

R_3

T_{11}	T_{12}	T_9	T_{10}
T_4	T_1	T_2	T_3
T_5	T_6	T_7	T_8

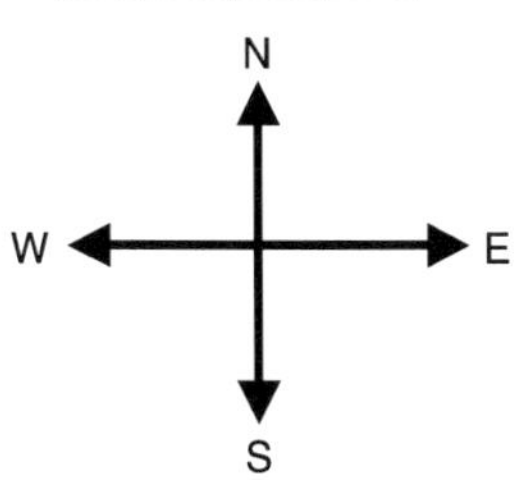

1. Layout: Factorial Randomized Block Design
2. Plot Size: 3 × 5 m
3. Treatments: 12
4. Replications: 3

Figure 5.1: Layout of the Experiment

Number of Effective Tillers

The number of the effective tiller of grasses were counted from 5 randomly selected plants in each treatment and averaged at 30 days and 60 days of sowing of legumes.

Leaf Length

The leaf length of the grasses were counted from 5 randomly selected plants in each treatment and averaged at 30 days and 60 days of sowing of the legumes.

Fresh Forage Yield

Crop cutting was taken by using 1 × 1 m quadrant and the green fodder weight was taken from each treatment at 60 days of sowing of the legumes. Then yield per hectare was calculated.

Dry Forage Yield

From each treatment, 1 kg. green fodder was taken and dried at the 60 days of the sowing of legumes. Then yield per hectare was calculated.

Finally the observations were put to statistical analysis.

Results

Results obtained from the field study of effect of phosphorus and legumes on the Trisankar pasture are presented.

Effect on the Grass (Trisankar *Pennisetum* hybrid)

Plant Population

Application of different levels of phosphorus and intercropping of legumes had effect on the plant population of the Trisankar (Table 5.1).

Plant Height

There was no significant difference in the plant height in 30 days of sowing of legumes in phosphorus levels alone. However in 60 days maximum plant height was observed at 90 kg phosphorus level in combination with stylo followed by 90 kg phosphorus with siratro intercropping (Table 5.1).

Table 5.1: Effect of Different Levels of Phosphatic Fertilizers and Pasture Legumes on the Plant Population Height (cm), Leaf Length (cm), Tussock Diameter (cm), Total Tillers (no) and Effective Tillers (no) of Trisankar Pennisetum Grass at 30 and 60 Days of Sowing of Legume

Treatments	*Plant Population (no)*		*Height (cm)*		*Leaf Length (cm)*		*Tussock Diameter (cm)*		*Total Tillers (no)*		*Effective Tillers (no)*	
	30	*60*	*30*	*60*	*30*	*60*	*30*	*60*	*30*	*60*	*30*	*60*
$G+P_0$	9	9	95.3	156.7	99.3	101.0	14.7	16.3	10.7	18.3	-	6.3
$G+P_{30}$	9	9	140.0	213.3	87.7	88.3	19.3	22.0	9.7	17.7	-	4.3
$G+P_{60}$	9	9	118.3	220.0	88.3	88.7	17.0	19.7	11.0	19.7	-	9.3
$G+P_{90}$	9	9	114.7	222.3	95.0	95.7	18.3	20.3	12.3	21.7	-	4.3
$G+L_a+P_0$	6	6	114.7	253.3	82.3	84.7	16.3	19.0	11.7	22.3	-	8.0
$G+L_a+P_{30}$	6	6	130.0	250.0	87.0	89.7	16.0	20.3	12.7	25.7	-	9.0
$G+L_a+P_{60}$	6	6	107.3	260.0	73.3	74.7	15.0	22.0	13.3	27.0	-	6.7
$G+L_a+P_{90}$	6	6	103.3	263.3	69.7	72.0	20.7	22.3	13.7	27.7	-	8.3
$G+L_b+P_0$	6	6	105.7	223.33	84.7	85.3	11.7	15.0	14.3	21.7	-	7.3
$G+L_b+P_{30}$	6	6	122.0	250.0	84.7	85.7	18.3	22.7	15.3	24.0	-	8.0
$G+L_b+P_{60}$	6	6	127.3	273.3	94.3	96.0	17.7	24.0	7.0	21.7	-	7.7
$G+L_b+P_{90}$	6	6	123.7	276.7	91.0	91.7	14.0	20.0	8.0	24.0	-	9.7

Contd...

Table 5.1–Contd..

Treatments	Plant Population (no)		Height (cm)		Leaf Length (cm)		Tussock Diameter (cm)		Total Tillers (no)		Effective Tillers (no)	
	30	*60*	*30*	*60*	*30*	*60*	*30*	*60*	*30*	*60*	*30*	*60*
CD at 5%												
GL	-	-	NS	18.72	10.5	7.3	NS	NS	NS	4.43	-	2.13
P	-	-	NS	21.61	NS	NS	NS	NS	NS	NS	-	NS
GL × P	-	-	NS	NS	NS	NS	NS	NS	NS	NS	-	NS
SE_m +												
GL	-	-	9.36	6.38	3.59	2.49	3.59	2.49	1.20	1.17	-	0.73
P	-	-	10.81	7.36	4.14	2.88	4.14	2.88	1.38	1.35	-	0.84
GL × P	-	-	18.73	12.76	7.14	4.98	7.14	4.98	2.4	2.34	-	1.45

G: Grass; P: Phosphorus at levels = 0,30,60,90 kg/ha; L_a: Legume a (*Stylosanthes hamata*); L_b : Legume b (*Macroptilium atropurpureum*).

Leaf Length

Application of different levels of phosphorus and intercropping of legumes showed more or less same effect on days of sowing (30 and 60 days) of legumes (Table 5.1).

Tussock Dimeter

Application of different levels of phosphorus and intercropping of legumes showed less effect on the tussock diameter of grass both at 30 and 60 days of sowing of legumes. In 30 days of sowing the highest tussock diameter was recorded (20.7cm) in 90 kg of phosphorus application with intercropping of stylo. However, in 60 days sowing the highest dimeter was recorded (24 cm) in 60 kg phosphorus level with intercropping of siratro (Table 5.1).

Total Tillers

Application of different level of phosphorus had a little effect on the tillering of the grass. However, with intercropping of stylo, the number of tillers increased with the phosphorus level. But intercropping of siratro showed a high number of tillers in 60 and 90 kg of phosphorus (Table 5.1).

Effective Tillers

Application of different levels of phosphorus had less effect on the flowering of the grass. However the intercropping of legumes showed more or less same number of effective tillers on different levels of phosphorus (Table 5.1)

Effect on the Legumes

Plant Population

More or less same number of plant population in two different days of sowing (30 and 60 days) at different levels of phosphorus. However, the intercropping of stylo recorded better performance than siratro at different levels of phosphorus (Table 5.2).

Plant Height

Different levels of phosphorus had little effect on stylo at both the days sowing. However, the siratro recorded a remarkable increase in height at both the days of sowing (Table 5.2).

Branching

Different levels of phosphorus had remarkable effect on stylo, whereas there was little effect of different levels of phosphorus on siratro (Table 5.2).

Table 5.2: Effect of Different Levels of Phosphatic Fertilizers on the Plant Population (nos), Height (cm), Branching (cm) of Pasture Legumes *Stylosanthes hamata* (stylo) and *Macroptilium atropurpureum* (Siratro) at 30 and 60 Days of Sowing of Legume

Treatments	*Plant Population (no)*		*Heights (cm)*		*Branching (cm)*	
	30	*60*	*30*	*60*	*30*	*60*
$G+L_a+P_0$	21.7	21.3	8.7	14.7	2.3	4.7
$G+L_a+P_{30}$	22.0	21.6	8.0	16.3	1.7	7.0
$G+L_a+P_{60}$	28.7	26.0	8.0	19.7	1.0	7.7
$G+L_a+P_{90}$	21.7	21.3	9.0	20.7	1.7	8.3
$G+L_b+P_0$	22.3	22.0	13.3	99.0	1.7	5.0
$G+L_b+P_{30}$	26.1	25.0	12.7	103.0	2.3	3.0
$G+L_b+P_{60}$	12.3	12.3	20.3	120.0	1.0	3.3
$G+L_b+P_{90}$	13.0	12.3	16.7	121.7	1.3	2.7
CD at 5%						
GL	7.85	6.2	3.59	19.33	0.42	1.78
P	NS	NS	NS	NS	0.49	NS
GL x P	NS	NS	NS	NS	NS	NS
SE_m+						
GL	2.68	2.11	1.20	6.60	0.14	0.61
P	3.09	2.44	1.41	7.61	0.17	0.70
GL x P	5.36	4.23	2.45	13.18	0.29	1.22

G: Grass; P: Phosphorus at levels = 0,30,60,90 kg/ha;
L_a: Legume a (*Stylosanthes hamata*);
L_b : Legume b (*Macroptilium atropurpureum*).

Effect on Forage Yield

Fresh Forage Yield

The fresh forage yield was highest (52.6mg ha^{-1}) in 90 kg phosphorus application. However, intercropping of stylo and siratro recorded 49.3 and 43.3 mg ha^{-1} respectively at 90 kg phosphorus level (Table 5.3).

Table 5.3: Dry and Fresh Forage Yield (mg ha^{-1}) of Trisankar, Trisankar-Stylo and Trisankar-Siratro Pasture

Treatments	*Fresh Forage Yield (mg ha^{-1})*			*Dry Forage Yield (mg ha^{-1})*		
	Grass	*Legume*	*Grass+ Legume*	*Grass*	*Legume*	*Grass+ Legume*
$G+P_0$	32.30	–	32.30	5.44	–	5.44
$G+P_{30}$	36.30	–	36.30	6.53	–	6.53
$G+P_{60}$	48.30	–	48.30	8.98	–	8.98
$G+P_{90}$	52.60	–	52.60	9.99	–	9.99
$G+L_a+P_0$	26.30	0.70	27.0	5.40	0.14	5.54
$G+L_a+P_{30}$	34.50	1.10	35.60	8.04	0.25	8.29
$G+L_a+P_{60}$	36.40	1.60	38.0	8.95	0.39	9.34
$G+L_a+P_{90}$	47.30	2.0	49.30	12.91	0.54	13.54
$G+L_b+P_0$	18.10	1.50	19.60	4.16	0.34	4.50
$G+L_b+P_{30}$	31.20	2.40	33.60	7.99	0.61	8.60
$G+L_b+P_{60}$	31.50	4.50	36.0	8.69	1.24	9.93
$G+L_b+P_{90}$	36.90	6.40	43.30	11.0	1.90	12.90

G: Grass; P: Phosphorus at levels 0,30,60,90 kg/ha;
L_a: Legume a (*Stylosanthes hamata*); L_b: legume b (*Macroptilium atropurpureum*).

Dry Forage Yield

The dry forage yield of grass was highest (9.99 mg ha^{-1}) in 90 kg phosphorus application. However, intercropping of stylo and siratro recorded 13.45 and 12.9 mg ha^{-1} respectively at 90kg of phosphorus application (Table 5.3).

Discussion

The experiment was laid out with the aim to know the effect of phosphatic fertilizer and pasture legumes on the Trisankar pasture. Results are being discussed in the following text.

Effect on Trisankar Grass

It was observed that leaf length, tussock dimeter and number of effective tillers in Trisankar grass was not significantly affected by the different level of phosphorus and intercropping of the pasture legumes. But the plant height was significantly affected. The plant height increased with increase in phosphorus levels along with associated legumes. The maximum height was recorded at 90 kg phosphorus level. However, the increase of height of plant was higher from 30 kg phosphorus to 60 kg phosphorus level. The response of phosphorus was higher in association with stylo compared to siratro. The number of tillers increased with increasing levels of phosphorus. The response was prominent in association with stylo at the three levels of phosphorus application. However, the association of siratro showed good response at 60 and 90 kg phosphorus levels.

Effect of Phosphatic Fertilizer on Pasture Legumes

The plant population of pasture legumes was affected very little by the different levels of phosphatic fertilizer. However, the plant height of two different legumes responded differently. The response was more in case of siratro than stylo. The reverse was the case with branching of the legumes.

Forage Yield of Trisankar Pasture

According to Prasad and Mukherjee (1984) the introduction of legumes in Dinanath grass increased the fresh forage yield over the control. In this experiment observation of the introduction of legumes, stylo and siratro by replacing the middle line of every three lines of grass, the fresh forage yield decreased compared to control. It was noted that the legumes could not compensate the loss of forage yield caused by the replacement of grass population as the Trisankar is a tall, tussock and profusely tillering grass.

But the dry forage yield increased over the control and showed linear response to the increasing levels of phosphorus in association with pasture legumes. The maximum dry forage yield 13.45 mg ha^{-1} was recorded at 90 kg phosphorus along with stylo followed by 12.9 mg ha^{-1} at 90 kg phosphorus along with siratro. Similar results of increased dry forage yield in the grass-legumes mixture over control was reported by Rai *et al.*, 1980, Velaydhan *et al.*, 1977.

It was also noticed that the fresh forage yield contributed by stylo was less than siratro. Similar trend was also seen in case of dry forage yield of the legumes. But in the mixed pasture the total fresh forage yield and dry forage yield at the 90 kg phosphorus level was highest in stylo mixed pasture than the siratro mixed pasture.

Conclusion

The effect of levels of phosphatic fertilizer with and without pasture legumes on the forage production of the Trisankar pasture was remarkable. Though the fresh forage yield was less in the grass + legumes + phosphorus pasture than the grass + phosphorus pasture, the dry forage was higher in grass + legumes + phosphorus pasture than grass + phosphorus pasture. This can possibly be ascribed to better nodulation of legumes due to the application of phosphatic fertilizer which ultimately resulted in the higher dry matter production.

Acknowledgement

The authors express their deep sense of gratitude to the Department of Animal Husbandry and Veterinary services, Government of Orissa for granting permission to carry out this research work as well as the Director, Indian Grassland and Fodder Research Institute, Jhansi (U.P.) for providing the necessary facilities to perform the field and laboratory experiments.

References

Bharucha FR (1858) The problems of grassland improvement in India. Curr. Sci. 61 (12): 600-660.

Burns W (1915) The improvement of natural grassland of India. 10: 288-293.

Dwivedi GK, Kanodia KC, Nirajan KP and Dixit OP (1985) Intercropping stylos with *Chrysopogon*. Indian J. Range Mgnt. 6(1 and 2): 59-64.

Kanodia KC and Patil BD (1991) Pasture establishment and management as a component of forage forestry. Indian J. Range Mgmt. 2 (1 and 2): 95-107.

Nirajan KP, Kanodia KC, Dwivedi GK and Rai P (1990) Comparative study of fertilizer and legumes and the forage production of *Chrysopogon fulvus* under rainfed condition for promotion of biomass in waste lands. Range Mgmt. and Agroforestry 11 (1): 21-24.

Prasad LK and Makherji SK (1984) Productivity of Dinanath grass in association with pasture legumes. Indian J. Range mgmt. 5: 39-42.

Rai P and Patil BD (1985) Production potential of *Stylosanthes* in pure and mixed swards. Indian J. Range Mgmt. 6: 35-37.

Rai P, Kanodia KC, Patil BD, Velayudhan KC and Agarwal R (1980) Nitrogen equivalence of range legumes introduced in natural grasslands. Indian J. Range Mgmt. 1(2): 97-101.

Shankaranarayan KA, Dabadghao PM and Kumar R (1975) Techniques for establishment of pasture legumes in grasslands of *Heteropogon*. Indian J. Agric. Sci. 45: 194-198.

Singh KA, Patil BD and Rai P (1984) Integration of forage food production through intercropping in semi-arid regions. Indian J. Agron 29: 510-515.

Velayudhan KC, Jayan PK, Rai P and Kanodia KC (1977) Technologies for establishment Townsville stylo *Stylosantnes humilis* in Sehima-Heteropogon grassland. Forage Res. 3: 43-48.

Chapter 6

Toxicity of Different Fungicides to the Rice-Field Cyanobacteria

Mihir K. Das*

P.G. Department of Botany, G.M. College (Autonomous), Sambalpur – 768 004, Orissa, India

ABSTRACT

The extensive use of pesticides has become imperative for achieving higher production, but at the same time, indiscriminate use of these pesticides is known to affect the cyanobacterial population in the rice fields. Cyanobacteria are known to contribute to the maintenance of soil fertility and are consequently considered to be an important component of soil biota. These organisms are more sensitive to fungicides than other groups of micro-organisms. It is therefore, essential that regular monitoring be undertaken to evaluate the influence of fungicides on cyanobacteria. In the present review, range of fungicide-induced biological responses, induction of biochemical changes in the organisms, the relative tolerance of cyanobacteria to fungicides etc. are highlighted.

Keywords: Cyanobacteria, Fungicides, Rice field.

* Tel.: 0663-2548197, (M) 9437393301; E-mail: mihir_gmc@rediff.com

Introduction

Rice is the staple food for at least half of the world's population. World-wise, it is harvested from approximately 146 million hectares of land (IRRI, 1991). The Asian rice growing area is 90 per cent of the global total, in a region which supports about 60 per cent of the world's population (IRRI, 1991). Thus, rice remains very much a crop of subsistence agriculture. The yield of rice is directly proportional to the nitrogen status of the soil. In most of the developing countries, the potential nitrogen requirement is often one-third or one-half of that of the developed nations and actual requirement is even less. In terms of nitrogen consumption per hectare of agricultural land, developing countries consume only an average of 8 kg N/ha (Venkataraman, 1981). This indicates the necessity of improving the soil nitrogen status. Although nitrogen occurs in the atmosphere in the form of molecular nitrogen in a concentration of 77 volume per cent, agriculture is in a state of poverty amidst plenty with regard to this nutrient.

Nitrogen fixation is a high energy budgeted process requiring an input of about 147 kcal to reduce one molecule of nitrogen and this is the reason why most of the higher plants have not evolved nitrogen fixing mechanisms; they simply may not be able to afford the energy. Blue-green algae (cyanobacteria) are the unique group of organisms which carry out both oxygen evolving photosynthesis and oxygen sensitive nitrogen fixation; the energy bill for the latter process being paid by the sun (Das and Adhikary, 2006). Rice is perhaps the only major cercal crop to whose nitrogen economy nitrogen-fixing cyanobacteria make a significant contribution (Roger and Kulasooriya, 1980).

Cyanobacteria

Cyanobacteria constitute the largest, most diverse and widely distributed group of photosynthetic prokaryotes. They have the ability to invade otherwise uninhabitable sections of the environment, namely, hot spring, ice-lands, volcanic soils and bare rocks where the combined nitrogen content is either absent or the temperature is in either of the extremes or a variety of natural or artificial chemicals are present (Whitton, 1992). In addition, these organisms multiply rapidly in the flood water in neutral to alkaline soils and release nitrogen slowly for meeting the nutrient requirement by rice (Tirol *et al.*, 1982, Whitton *et al.*, 1988). Agronomic potential of cyanobacteria in the rice cultivation was first recognized by De (1939), who attributed the natural fertility of tropical rice fields to biological nitrogen fixation by these organisms. Since then, long term fertility experiments and nitrogen fixation measurements have confirmed the importance of cyanobacteria in maintaining a moderate, but constant rice production in fields receiving no fertilizer (Roger, 1988).

Pesticide

Pesticide as an Integral Part of Agriculture

The increased use of fertilizers has become an integral part of the cultivation practices. With higher fertilizer inputs, disease and pest problems have become prominent which endangered the yield necessitating the large scale use of pesticides. Until 1850s, the use of pesticides against pests was not in practice. Before 1940, they were used in relatively small amount and considered as 'safe pesticides' which were derivatives of various plants. But from 1940 onward, the development of numerous synthetic organic compounds or so called 'second generation pesticides' completely changed the basis and strategies of pest control. Thus, during 1950s and 1960s, huge amount of organic pesticides were widely used in agriculture and horticulture to facilitate the production and protection of food, feed and fibre in greater quantity. During 1960s, it was realized that any one single pesticide did not completely eliminate all the species of a target pest. This created a number of trends in use and development of new pesticides for effective control. As a result, the number of chemical pesticides has increased during the last past 50 years. The pesticides included a number of insecticides, fungicides, herbicides, nematocides etc.

Pesticide Hazard

Though pesticides were considered as the miracle drugs for the plants during 1950s, doubts were raised concerning the hazardous effects of these chemicals on the environment from the early 1960s. The DDT used in the field has been detected in the mother's milk. The travel of DDT and other pesticides through various components of the food chain has not only affected the flora and fauna, but also, has reappeared in the consumable food products. Richael Carson was one of the first to become aware of the danger of pesticide pollution and in her famous book entitled 'Silent Spring', she produced evidences to toxic effects of pesticides upon wildlife and potential hazards to man (Misra and Mani, 1994). Despite increasing research efforts towards crop improvement by methods obviating the use of pesticides (Day, 1987), agriculture remains heavily dependent on these agrochemicals, (Gadkarni, 1988). The soil, being the ultimate sink in the rice fields, harbours many beneficial cyanobacteria, which are potentially susceptible to pesticides. Cyanobacteria are known to contribute to maintenance of soil fertility (Singh, 1961; Venkataraman, 1972). It is therefore, imperative that regular monitoring be undertaken to evaluate the influence of pesticides on these organisms.

The pesticide toxicity is regarded only as the lethal effects of the chemicals upon a target organism. It directly changes the growth rate as well as metabolic activities such as photosynthesis and respiration. It may have a toxic effect upon certrain groups of microorganisms (Padhy, 1985). The effect of this kind may alter population equilibrium in the indirect manner by changing the competitive efficiency of one or other group. Pesticide may promote either directly or indirectly, the growth of one or more type of organisms, which could be either beneficial or undesirable in paddy field ecosystem. Finally, toxic action of pesticides causing death of certain groups of organisms would result in the release of nutrients by the decay of these organisms, thus inducing a corresponding increase in the number of pesticide tolerant forms.

The research work so far carried out in various laboratories of the world relates to the interaction of pesticides to cyanobacteria through *in vitro* determinations and/or under *in vivo* conditions (field condition). Most of the works pertain to the determinations of lethal doses of pesticides on different species/strains of cyanobacteria. The more recent studies have been more biochemical and physiological in nature than the earlier ones, departing from examination of pesticide-induced qualitative and quantitative fluctuations in soil microbial populations and concentrating more on factors such as the impairment of nitrogen fixation (Kapoor and Sharma, 1980; Kaushik and Venkataraman, 1983). Although studies in the interaction of cyanobacteria with insecticides continue to be widely conducted, there has been an increase in the amount of attention paid to fungicides and herbicides. Since 1979, emphasis has also been laid on the examination of the response of cyanobacteria to mixtures of pesticides (Stratton, 1983; Tandon *et al.*, 1988). This is encouraging, since in rare cases, would any pesticide be present in isolation in the soil.

Fungicides

The extensive use of fungicides has greatly improved many facets of agriculture, but at the same time has created many unintended problems. The effect of fungicides on non-target organisms, either direct or indirect, is a significant problem. Cyanobacteria are very important groups of soil biota and interference of the activity of algae by soil incorporated fungicides may be expected to have potentially serious consequences on the overall productivity of soil. Cyanobacteria are more sensitive to fungicides than other groups of micro-organisms. In the present review effect of fungicides on growth and biological nitrogen fixation, mode of action and factors influencing the fungicides and algal interaction have been discussed.

Effect of dichlone on cyanobacteria was first reported by Fitzerald *et al.* (1952). *Microcystis aeruginosa* blooms in ponds and non-bloom forming cyanobacteria were cleared by 30 to 55 ppm and 5 ppm dichlone respectively. Fitzerald and Skoog (1954) conducted large scale field trials in clearing cyanobacteria with dichlone (10 to 100 ppb). Dichlone at 10 ppb prevented the growth of *Anabaena* sp. (Bisiach, 1970). The growth of *Anacystis nidulans* was prevented by 140 ppb dichlone (Whitton and McArthur, 1967).

Dithiocarbamate fungicides are the oldest one and many of the members are still actively used to control the plant diseases with greater effectiveness. The members include zineb, thiram, ziram, maneb, moncozeb etc. They are being used in soil treatment, seed application and foliar application. Zineb (zinc ethylene bisdithiocarbamate) was found highly toxic to *Anabaena* and *Nostoc* at very low concentrations, however, some other strains tolerated even up to 50 ppm. The growth of *Cylindrospermum* sp and *Nostoc muscorum* was inhibited at 1 and 5 ppm of zineb respectively in nitrogen deficient medium, but tolerated higher concentrations in nitrogen supplemented medium (Venkataraman and Rajyalakshmi, 1971). They further reported that zineb at 100 kg/ha did not inhibit the growth of eight strains of *Anabaena,* nine (out of 16) strains of *Nostoc, Aulosira fertilissima, Tolypothrix tenuis* and *Anacystis nidulans.*

Zineb is found to have a sub-lethal range of 1 to 15 ppm with *Cylindrospermum* and 5 to 25 ppm with *Nostoc muscorum* in C-N medium. The complete toxic levels in this medium were 25 and 300 ppm for *Cylindrospermum* and *N. muscorum* respectively. In C+N medium, the sub-lethal range was 5 to 20 ppm and 10 to 30 ppm in *Cylindrospermum* and *N. muscorum* respectively, the complete lethal ranges correspondingly were 30 and 35 ppm zineb (Panigrahy, 1984). Panigrahi (1984) reported that *Cylindrospermum* and *Nostoc* tolerated zinc *in vitro* at 5 and 15 ppm respectively. Schluter (1964) reported little effect of zineb on the growth of *Scenedesmus quadricauda* and *Kirchneriella lunaris.*

Ziram is another member of dithiocarbamates being used for the control of foliar diseases. Thirty-two species of cyanobacteria were treated with several doses of ziram. Diatoms were more sensitive as compared to cyanobacteria and green-algae (Maloney and Plamer, 1956). But this fungicide had a stimulatory effect on growth of *Cylindrospermum* sp. and *Nostoc muscorum* at the 0.01 ppm level in both C+N and C-N media. The sub-lethal range for *Cylindrospermum* was found to be 0.05 to 0.13 ppm, whereas for *N. muscorum,* this range was from 0.1 to 0.4 ppm in C-N medium. The complete lethal dose in C-N medium was 0.1 ppm for *Cylindrospermum* and 0.5 ppm for *N. muscorum.* The sub-lethal range in C+N medium was 0.05 to 0.13 ppm for *Cylindrospermum* and 0.1 to 0.5 ppm for *N. muscorum.* The complete lethal doses in C+N medium were 0.02 and 0.6 ppm for *Cylindrospermum* and *N. muscorum* respectively (Panigrahy, 1984).

Mancozeb (Zinc ethylene bisdithiocarbamate and manganese ions) is still another dithiocarbamate fungicide, highly effective in the control of foliar diseases. It is being intensively used where long term effect is being required. Mancozeb in the range of 5 to 100 ppm was least toxic to 27 strains of cyanobacteria (Venkataraman and Rajyalakshmi, 1972). This fungicide had a sub-lethal range up to 7 ppm for *Cylindrospermum* in both C-N and C+N media and the level of 9 ppm was completely toxic in both the media. The sub-lethal range of mancozeb was up to 10 ppm with *N. muscorum* in both C-N and C+N media and the level of 15 ppm was completely toxic in both media. The highest tolerance level in both the media (C-N and C+N) was 5 ppm mancozeb for both the strains on agar plates (Panigrahy, 1984). Seed treatment fungicide, thiram (tetramethyl thiuramdi-sulphide) was very toxic as it completely removed *Nostoc* and *Anabaena* at the rate 1.2-12 ppm; however, less than 1.2 ppm, the growth was stimulated by 200 per cent (Cameron and Julian, 1984).

Virtually all data were collected under *in vitro* conditions, except a few, in which the benzimidazole carbendazim at 2.9 ppm, the recommended application rate, was applied to soil held in glass cylinders in the laboratory. The fungicide affected neither the growth nor nitrogenase activity of the soil algal and cyanobacterial flora. Gangawane (1980) also observed that carbendazim (as Bavistin®) could be tolerated by nitrogen-fixing cyanobacteria. This likewise applied to the chlorophenyl fungicide quintozene (as Brassicol®). In a separate study, tolearance levels of carbendazim (as MBC) were 500 ppm for *Calothrix* sp. and *Nostoc* sp. and 1,000 ppm for *Aulosira fertilissima* and *Tolypothrix tenuis, Westiellopsis prolifica,* however, was intolerant of concentrations as low as 1 ppm (Gangawane and Saler, 1979). Shukla *et al.* (1990) noted the inhibition of growth of unspecified algae of a standy loam soil following treatment with carbendazim (as Bavistin®) at 0.25, 0.5, 0.75 and 1 per cent. Other fungicides (copper oxychloride, as Blitox; EBP, as Kitazin®, mancozeb, as manzeb and Dithane ®) used in the same study and at the same concentrations also inhibited algal growth. The toxicity of these fungicides can be summerized as follows (most to least toxic: Kitazin® > Dithane® > Blitox® > Bavistin® > Manzeb (Shukla *et al,* 1990).

Nabam at 1 ppm affected photoassimilation of acetate of *Nostoc muscorum* (Moore, 1967). Captan at 1 ppm level suppressed development but not growth of *N. muscorum* (Moore, 1967). Growth of *Anacystis nidulans* was inhibited by a low concentrations of captan (Moore and Dorward, 1968). Phthalimide fungicides examined by Gangawane and Saler (1979) were found to be relatively non- toxic to *Calothrix* sp., *Nostoc* sp.. *Aulosira fertilissima, Tolypothrix tenuis* and *Westiellopsis prolifica;* capton (as Hexacap®) was tolerated by all five organisms at upto 500 ppm active ingredient and captafol (as Difolatan®) was tolerated at 100, 300, 300, 500 and 1000 ppm active ingredient by *Calothrix sp., Aulosira fertilissima, Westiellopsis prolifica, Tolypothrix tenuis* and *Nostoc* sp. respectively (Gangawane and Saler, 1979). The growth of *Aulosira fertilissima* was actually stimulated by captafol. The authors attributed this to active metabolism and adaptation to the fungicide. The low toxicity of captan to cyanobacteria observed by Gangawane and Saler (1979) was not recorded by Cameron and Julian (1984), in spite of the fact that both studies employed liquid cultures with growth measurement by pigment extraction. In fact, treatments of greater than 15-24 ppm of captan and also other fungicides (benomyl, carboxin and thiram) were lethal to *Anabaena* sp., *Lyngbya* sp., *Nostoc* sp. and *Mastigocladus laminosus*. These concentrations are significantly lower than recommended field application rates (Cameron and Julian, 1984). Of the four fungicides examined in this study, thiram was the most and benomyl, the least toxic. Low concentrations of each of the four stimulated the growth of all test cyanobacteria, due possibly to the control of fungi or other heterotrophs that may have been present as contaminants (Cameron and Julian, 1984).

Hutber *et al.* (1979), while working with thiram and the pyrimidinol fungicide ethirimol (as Milstem®), observed that their toxicity to the cyanobacteria, *Aphanocapsa* sp, *Anabaena variabilis* and *Nostoc* sp. was intermediate in a range of other pesticides tested in the same study. These were completely inhibitory at 100 ppm. Sharma and Gaur (1981) suggested that a reduction in nitrogen fixation in strains adapted to the mercurial seed dressings PMA (Phenylmercury acetate) and MEMC (2-methoxy ethylmercury chloride, as Agallol®) may have been caused by the inhibition of some step(s) in nitrogen-fixation and/or protein synthesis. Tolerance limits of the test cyanobacteria to these organomercury fungicides were much lower (up to 0.7 ppm PMA and 2 ppm MEMC) than to the insecticides lindane and leptophos (up to 15 and 300 ppm, respectively) in the same study.

Ceresan did not inhibit growth of 8 strains of *Anabaena* and 11 strains of *Nostoc* and *Aulosira fertilissima* at 20 kg/ha and of *Tolypothrix tenuis* and *Anacystis nidulans* at 200 kg/ha. Thus the latter two genera are more resistant than the former three genera (Venkataraman and Rajyalakshmi, 1971). They also reported that among 10 *Anabaena* strains tested for their resistance to ceresan, nine could tolerate 10 ppm, but one strain was inhibited by levels higher than 1 ppm (Venkataraman and Rajyalakshmi, 1972).

Panacide (dichlorophene) at 20 ppm was toxic to *A. fertilissima* and *Nostoc* (Gupta and Saxena, 1974). *Westiellopsis* sp., *Aulosira* sp., *Nostoc* sp., *Tolypothrix* sp. and *Calothrix* sp. tolerated the highest concentration (1000 ppm) of brassicol. There were 17.1 and 10.2 per cent decrease of growth over control in *Aulosira* sp. and *Calothrix* sp. respectively, at 500 ppm brassicol. In *Westiellopsis* sp. and *Nostoc* sp., brassicol at 1000 ppm caused 48.3 and 16.9 per cent reduction in growth over the control. *Tolypothrix* was not affected by 1,000 ppm brassicol (Gangawane, 1980). Bavistin was toxic to *Calothrix* sp. at 1 ppm level. This fungicide at 50 ppm caused 5.8 per cent increase in growth of *Tolypothrix* sp. and at 100 ppm 5.6 per cent increase in growth of *Nostoc* sp. growth of *Aulosira* sp. was decreased by 21.4 per cent by the level of 100 ppm bavistin. The level of 300 ppm bavistin caused 31.6 per cent reduction in growth of *Westiellopsis* sp. Fytolan was not toxic to *Aulosira* sp. at 100 ppm level, where as this concentration caused 58.3 and 39.7 per cent decrease in growth of *Westiellopsis* sp. and *Calothrix* sp. respectively. Growth of *Nostoc* sp. and *Tolypothrix* sp. was reduced by 18.8 per cent and 6.25 per cent by a 500 ppm level of fytolan (Gangawane, 1980).

The introduction of systemic fungicides revolutionized chemical control of plant diseases. They have several attributes, such as they are absorbed by the plant parts, translocate, persist in the plant and control pathogens away from the place of application. Further, they are required in very small quantity as compared to non-systematic fungicides (Vyas and Shroff, 1992). They are used as seed and soil treatment and also foliar and fruit application. Among them, MBC (methyl benzimidazole carbamate) a degradation product of carbendazim and benzimidazoles was found to be highly toxic to *Calothrix* sp., only 1 ppm reduced the growth of algae considerably (Gangawane, 1980). The fungicide at 50 ppm increased the growth of *Tolypothrix* sp. by 5.8 per cent and similar stimulation occurred at 100 ppm in *Nostoc*. The growth of *Aulosira* was decreased by 21.4 per cent at 100 ppm of carbendazim and a dose of 300 ppm decreased it by 31.6 per cent in *Westiellopsis* sp. Thiophanate methyl (Topsin-M) is a member of benzimidazole group, which caused 11.5 per cent reduction in the growth of *Calothrix* at 100 ppm and at similar concentration, *Westiellopsis* was reduced by 31.6 per cent (Gangawane and Kulkarni, 1979). They further observed that in *Aulosira* and *Tolypothrix* at 300 ppm thiophanate methyl reduced the growth by 12.8 and 14 per cent, respectively and at 1000 ppm, it reduced the growth of *Nostoc* sp. by 22.5 per cent. MBC decreased the growth of *Nostoc* and *Calothrix* by 7.9 and 1.0 per cent respectively. *Westiellipsis prolifica* could not tolerate MBC while *Aulosira fertilissima* and *T. tenuis* had little effect at 1000 ppm; *Nostoc* and *Calothrix* were unaffected by 500 ppm of MBC (Gangaware and Fater, 1979).

Hosmani (1992) studied the interaction effects of two fungicides (Kitazin and Baytan) on the heterocyst frequency of *Calothrix membranaceae*. Single application of kitazin at various concentrations indicate that heterocyst frequency decreases with increase in the fungicide concentrations. However, the combined effect differs considerably and expected values indicate that the combinations at 0.0001 ppm is synergistic, while at 0.001 ppm, the effect is additive. This indicates that either 0.0001 ppm or 0.001 ppm has a beneficial effect on the heterocyst frequency. In the other two concentrations the antagonism increases with the increase of the concentrations of the fungicides. Dikshit and Tiwari (1992) observed stimulation in *Anabaena variabilis* by benlate at 100 ppm and by dithane at 100-200 ppm concentrations, while in *Aulosira implexa* var. *tenuis* by 13-bavistin at 100-200 ppm concentrations. Lethal effect was observed in *Anabaena variabilis* at 100 ppm of captan, 400 ppm of cyathion; in *Aulosira implexa* var. *tenuis* at 100 ppm of captan, 200 ppm of dithane and cyathion; in *Aulosira fertilissima* var. *tenuis* at 400 ppm of dithane, 200 ppm of cyathion; in *Scytonema chiastum* at 100 ppm of 13-bavistin, 400 ppm of cyathion and in *Scytonema stuposum* at 400 ppm of bavistin, benlate and dithane, 200 ppm concentrations of captan and cyathion.

Anand and Subramanian (1997) studied the physiological response of *Nostoc calcicola* to fungicides, bavistin and dithane. The alga showed better growth at 1 ppm of both the fungicides and chlorophyll-α was more than that obtained for cells grown without any amendment by BG-11-N medium. Nitrogenase activity was not suppressed completely, rather, dithane showed 65 per cent activity. In dithane treated cultures, the photosynthetic activity was more than with bavistin, but the unamended BG-11-N cultures showed highest activity. The CO_2 fixation was 9.4 moles in dithane and 9.0 moles CO_2 fixed/h/mg chlorophyll-α in bavistin. Total nitrogen, carbohydrate and protein content were maximum in dithane.

Mechanism of Action of Fungicides

Fungicides are *per ce* toxic to the algal cells and filaments. The fungicides reduced the cell number, affect chlorophyll concentration and responsible for their destruction, decrease oxygen production, check carbon dioxide fixation and caused an increase in the proportion of labeled carbon in sugar and glycine, accompanied by a reduction in labelled carbon in lipids and glutamic acid (Zweig *et al*, 1968). Rapid and irreversible alterations in cell permeability were observed. Ultra structural changes have also been observed. Usually fungicides altered cell membrane permeability and hence the primary site of fungicide attacks is cell membrane which might change the configuration of sulphydryl group (Vyas and Shroff, 1992).

Conclusion

Cyanobacteria are unique group of micro-organism of ecosystem and thus application of fungicides drastically altered their activity. It was observed that most of the studies of effect of fungicides on cyanobacteria are carried out in the laboratory. The effect of fungicides on algae in soil are considerably different from effects on the same algae in laboratory condition. Even of the fungicides are only 1 per cent effective in soil as in pure culture, they could still have dramatic effects on soil nutrition. The environmental factors also affect the chemical toxicity. The above findings clearly indicate the involvement of experimental conditions in regulating the intensity of the observed response of the test organism and further provide evidence of the extreme complexity of the interactions between the fungicides and cyanobacteria in the soil system.

Acknowledgement

The author expresses his deepest sense of gratitude to Prof. S.P. Adhikary, P.G. Department of Botany, Utkal University, Vani Vihar, Bhubaneswar for his constant encouragement and valuable suggestion during the preparation of this manuscript.

References

Anand N and Subramanian TD (1997) Effect of certain pesticides on the physiology of *Nostoc calcicola*. Phykos 36 (1 and 2): 15-20.

Bisiach M (1970) Algal infestation in Italian soils. Riso 19:129.

Cameron HJ and Julian GR (1984) The effects of four commonly used fungicides on the growth of cyanobacteria. Plant Soil 78:409-415.

Das MK and Adhikary SP (2006) Toxicity of different insecticides to rice field cyanobacteria. In: Glimpses of Cyanobacteria. Ed.

Gupta RK, Kumar M and Paliwal GS. Daya Publishing House New Delhi, PP. 122-134.

Day PR (1987) Crop improvement: constraints and challenges. In: British crop protection Conference on weeds. BCPC Pub. Surrey, UK, PP.3-12.

De PK (1939) The role of blue-green algae in nitrogen fixation in rice fields. Proc Roy Soc London 127: 121-134.

Dikshit G and Tiwari GL (1992) effect of pesticides on certain nitrogen-fixing cyanobacteria. In: Proc. NaH Symp Cyanobacterial Nitrogen Fixation. Ed. Kaushik BD; Ass Publ Co, New Delhi, PP 495-500.

Fitzgerald GP and Skoog F (1954) Control of blue green algae blooms with 2,3- dichloronaphthoquinone. Sewage Ind Wastes 26:1136.

Fitzgerald GP, Gerloff GC and Skoog F (1952) Studies on chemicals with selective toxicity to blue green algae. Sewage Ind Wastes 24:888.

Gadkari D (1988) Assessment of the effects of photosynthesis inhibiting herbicides diuron, DCMU, metamitron and metribuzin on growth and nitrogenase activity of *Nostoc muscorum* and a new cyanobacterial isolate strain G4. Bio Fertile Soils 6:50-54.

Gangawane LV (1980) Tolerance of nitrogen–fixing blue green algae to brassicol, bavistin and fytolan. J Ind Bot Sci 59: 157-160.

Gangawane LV and Kulkarni L (1979) Tolerance of certain fungicides by nitrogen fixing blue green algae and their side effects on rice cultivars. Pesticides 13:37.

Gangawane LV and Saler RS(1979) Tolerance of certain pesticides by nitrogen-fixing blue green algae. Curr Sci 48:306-308.

Gupta GS and Saxena PN (1974) Effects of panacid on some green and blue green algae. Curr Sci 43:492.

Hosmani SP (1992) Interaction of two fungicides (Kitazin and Baytan) on the heterocyst frequency of *Calothrix membranacea* Schmidle. In: Proc Natl Symp cyanobacterial nitrogen fixation. Ed. Kaushik BD. Ass Publ Co, New Delhi PP 535-537.

Hutber GN. Rogers LJ and Smith AJ (1979) Influence of pesticides on the growth of Cyanobacteria. Z Allg Microbiol 19: 397-402.

IRRI (1991) World Rice Statistics, Los Banos, Philippines.

Kapoor K and Sharma VK (1980) Effect of certain herbicides on survival, growth and nitrogen fixation of blue-green alga *Anabaena doliolum* Bharadwaja. Z Allg Microbiol 20: 465–469.

Kaushik BD and Venkataraman GS (1983) Response of cyanobacterial nitrogen fixation to insecticides. Curr Sci 52: 321-323.

Maloney TE and Palmer CM (1956) Toxicity of six chemical compounds to thirty cultures of algae. Water Sewage Works 103: 509.

Misra SG and Mani D (1994) Agricultural Pollution Vol. II. Ashish Publ House, New Delhi 11 P.

Moore RB (1967) Algae as biological indicators of pesticides. J Phycol 3:4.

Moore RB and Dorward DA (1968) Accumulation and metabolism of pesticides by algae. J Phycol 4: 7.

Padhy RN (1985) Cyanobacteria and Pesticides. Residue Reviews. 95: 1–44.

Panigrahy KC (1984) Physiological and genetical effect of pesticides on blue-green algae: effect of carbamate pesticides. Ph.D. Thesis, Berhampur University (Orissa) India.

Roger PA (1988) Reconsidering the utilization of blue-green algae in wetland rice cultivation. In: Biological nitrogen fixation associated with rice production. Eds Dutt SK and Sloger C. Oxford and IBH Publ Co Pvt Ltd. New Delhi, PP. 119–137.

Roger PA and Kulasooriya SA (1980) Blue-green algae and Rice IRRI, Manila, Los Banos.

Schluter M (1964 b) Erste Untersuchungsergebnisse Uber die algizide wirkung funf verschiedener Fungizide Chlorophycean gegenuber. Deut Fischerel–Ztg. 12: 373.

Sharma VK and Gaur YS (1981) Nitrogen fixation by pesticide–adapted strains of paddy field cyanophytes. Int J Ecol Environ Sci 7: 117–122.

Shukla DS, Chaturvedi R and Srivastava SK (1990) Effect of various fungicides on microbial population of soil and some soil factors. J Environ Biol 11: 291–298.

Singh RN (1961) *Role of blue-green algae in nitrogen economy of Indian agriculture*, ICAR, New Delhi.

Stratton GW (1983) Interaction effects of permethrin and atrazine combinations towards several non-target micro-organisms. Bull Environ Contam Toxicol 31: 297–303.

Tandon RS, Lal R and Rao VVSN (1988) Interaction of endosulfan and malathion with blue-green algae *Anabaena* and *Aulosira fertilissima*. Environ Poll 52: 1–9.

Tirol AC, Roger PA and Watanabe I (1982) Fate of nitrogen from a blue-green alga in a flooded rice soil. Sci Plant Nutr 28: 559–569.

Venkataraman GS (1972) *Algal biofertilizer and rice cultivation*. Today and Tomorrow's Printers and Publishers, New Delhi.

Venkataraman GS (1981) Blue-green algae: a possible remedy to nitrogen scarcity. Curr Sci 50: 253–256.

Venkataraman GS and Rajyalakshmi B (1971) Tolerance of blue-green algae to pesticides. Curr Sci 40: 143–144.

Venkataraman GS and Rajyalakshmi B (1972) Relative tolerance of nitrogen–fixing blue-green algae to pesticides. Ind J Agric Sci 42: 119-121.

Vyas SC and Shroff VN (1992) Effect of fungicides biofertilizer agents: 1. Cyanobacteria. In: Proc Natl Symp Cyanobacterial nitrogen fixation. Ed. Kaushik BD. Ass Publ Co, New Delhi, PP 287–292.

Whitton BA (1992) Diversity, Ecology and Taxonomy of the cyanobacteria. In: *Photosynthetic Prokaryotes*. Ed. Mann. NH and Carr NG, Plenum Press, New York, PP 1–51.

Whitton BA and McArthur K (1967) The action of two toxic quinones on *Anacystis nidulans*. Arch Mikrobiol 57: 147.

Whitton BA, Aziz A, Francis P, Rother JA, Simon JW and Tahmida ZN (1988) Ecology of deep water rice fields in Bangladesh. I. *Physical and chemical environment*. Hydrobiol 169: 3–67.

Zweig G, Hitt JD and Mc Mohan R (1968) Effect of certain quinones, diquat and diuron on *Chlorella pyrenoidosa* Chick (Emerson Strain). Weed Sci 16: 69–73.

Chapter 7

Environmental Impact of Pesticides: A Review

B.B. Mishra*

Minor Irrigation and Water Use Training Institute, Bhubaneswar, Orissa

ABSTRACT

Growing concern is being expressed over the adverse effect of the pesticides used indiscriminately in agriculture all over the globe. In an agriculture-based country like India, the problem of pesticide residues in the foodstuffs of both, plant as well as animal origin has already assumed an alarming proportion. Pollution of soil, air, water, development of resistance in the target pests to pesticides, pest resurgence causing minor pests to emerge as major ones and destruction of a variety of beneficial insects, useful soil microbes and predators and parasitoids that help in the natural suppression of pests are among other drastic consequences of the over-use of pesticides having serious environmental implications. The problem essentially calls for an integrated and ecofriendly approach in pest management rather than depending solely on chemical pesticides in order to minimize the pesticide load on the environment. The integrated pest management employs a variety of pest management methods and techniques e.g. physical, cultural, botanical, biological, biotechnological and

* Corresponding Address: Plot No. 203, Lane 8, Jagannath Vihar, Baramunda, Bhubaneswar. Ph: 2550323.

chemical ones combined together, so that the pest population can be maintained below a level causing economic injury to the crops without causing damage to the ecosystem. This can go a long way in contributing to food and nutritional security while maintaining the ecological balance at a reasonably safe level.

Keywords: Pesticide, Food, Soil, Water.

Introduction

The fact that agricultural production to a large extent depends on prevention of crop losses caused by crop pests has given rise to an exponential increase in the production and consumption of chemical pesticides over the past three to four decades. The total amount of pesticides used in the country increased from 154 metric tonnes in 1953-54 to nearly 84,000 metric tonnes in 1993-94 with a corresponding increase in the per hectare consumption from 1.2 g ha^{-1} in 1953-54 to as much as 431 g ha^{-1} in 1992-93. However, there has been a trend of decline in consumption of pesticides in India in all the states in late nineties. At present India is the largest producer of pesticides in South Asia with a production of 88,751 metric tonnes (1998-99). The total installed capacity is about 124, 000 metric tonnes for the manufacture of 62 technical grade of pesticides, which meet 95 per cent of the total need of the country. India is the third largest consumer of pesticides in the world and highest among the South Asian countries. The pesticides used in agriculture broadly include insecticides, fungicides, herbicides, nematicides, rodenticides, avicides, bactericides, molluscicides and ascaricides etc. Upto 1995-96 the major group of chemicals used in agriculture was insecticides (80 per cent) followed by fungicides (10 per cent), herbicides (7 per cent) and others (3 per cent). Thereafter, the consumption of insecticides declined with simultaneous increase in the consumption of herbicides and fungicides. The consumption of insecticides in 1999-2000 was 60 per cent, fungicides 21 per cent, herbicides 14 per cent and others 5 per cent. Although benefits rendered by pesticides to safeguard crop production by curtailing the crop damage due to pests is worth appreciating, the deleterious effect of these toxic chemicals on the environment is phenomenal. According to a study, only 1 per cent of the total pesticide applied is effective in controlling pests where as remaining 99 per cent goes into various environmental systems (Dhaliwal *et al*, 2001). The pesticide residue in the food chain has already proved to be a potential source of danger not only to the human being but also to the domestic animals, birds, fishes and non-target fauna of the ecosystem. Moreover, there are ample evidences that the pesticides also drastically affects the ecosystem through destruction of predators and other non-target beneficial insects, development of resistance in the target pests to pesticides and pest resurgence causing minor pests to emerge as major ones which in turn further necessitates higher consumption of pesticides resulting in the phenomenon of "pesticide treadmill". Although the actual estimate of damage caused by pesticides in India is not available, it is assumed that every one-dollar worth of pesticide used causes an estimated damage of 5-10 dollars to the environment and public health alone (Dhaliwal *et al.*, 2001).

Effect of Pesticides in Food

A number of trials conducted across the country have revealed the existence of pesticide residues in a wide range of food items including cereals, pulses, oilseeds, vegetables, spices, fruits, honey, milk, and other animal products (Table 7.1). Ironically enough, even human milk has also been reported to contain pesticide residues. The residues at bioconcentrations can cause serious toxic hazards to the consumers (Awasthi, 1985), as even small quantities of these residues ingested daily along with food can build up a high level in the body fat (Subba Rao, 1999). The market samples surveyed and analyzed in India over years from 1965 to 1998 revealed that as much as 59 per cent of them were contaminated with pesticides, 20 per cent of which contained pesticides above the maximum residue limit (Kathpal *et al.*, 2000).

The pesticides, which have been commonly detected in the foodstuffs, include HCH, DDT, monocrotophos, cypermethrin, quinalphos, aldrin and endosulphan etc. (Table 7.2). Unfortunately the milk and milk products serve as the major source of dietary intake of DDT followed by oils and fats in both vegetarian and non-vegetarian diets. The higher dietary intake of DDT residues through vegetarian diet as compared to the non-vegetarian is attributed to greater contribution of cheese in the diet (Subba Rao, 1999). The dietary intake of HCH is also observed chiefly through milk and milk products followed by meats and eggs for the non-vegetarian diet, whereas in case of vegetarian diet it is through cereals followed by milk and milk products (Kalra *et al.*, 1983). Singh *et al.* (1993) have presented an estimate of daily intake of DDT and HCH through consumption of milk contaminated at mean level in different states of India (Table 7.3). In contrast to chlorinated hydrocarbons, the organophosphates have not shown such residual effects. The non-persistence of organophosphate compounds in milk of cows is supposed to be due to their inactivation by rumen fluid of the cow (Lodha *et al.,* 1989). Due to the ubiquitous presence of DDT and HCH residues in all food commodities it has been estimated that the dietary intake of these pesticides often exceeded the acceptable daily intake (ADI) in India (Singh *et al.,* 1994). After the ban on DDT and HCH was imposed, the residues of these pesticides have however started showing a declining trend. But lindane residues have still been found to exceed the prescribed MRL of 0.01mg kg l^{-1} in 23 out of 24 samples analyzed during 1999-2000 (Dhaliwal *et al.,* 2001). In case of vegetables, a number of samples collected and tested have been reported to contain the residues of some commonly used organophosphorus pesticides like monocrotophos, quinalphos, triazophos etc. above the prescribed MRL (Dhaliwal *et al.,* 2000b). Recent report of the All India Network Project on Pesticide Residues (ICAR) has also revealed the persistence of residues of different pesticide in vegetables, fruits, spices, fish, cotton and butter etc. (Table 7.4). However the residues above MRL have been reported only in a small percentage of the samples analyzed.

Table 7.1: Pesticide Contamination of Different Foodstuffs in India

Foodstuff	*Number of Samples Analyzed*	*Number of Samples Found Contaminated with Pesticide Residues*
Wheat	1352	628
Rice	463	405
Pulses	487	211
Oils and oilseeds	588	488
Vegetables	6803	3642
Fruits	458	192
Spices	284	192
Animal products	506	323
Milk	1588	1358
Honey	148	135

Source: Kathpal and Kumari (2000).

Table 7.2: Pesticide Residues Identified in Various Food/Feed Items

Pesticide	*Food/Feed Items having Toxic Residues*
HCH	Rice, rice bran, paddy straw, tomato, carrot, cucumber, onion, radish, brinjal, okra, bitter gourd, green chilies, dry chilies, chilly powder, water samples, certain brands of cooking oils, milk products like milk powder, butter, cheese, infant milk powder and egg.
DDT	Cooking oil, milk, butter, maize, fish, egg, Livestock feed, human breast milk. Monocrotophos Tomato, onion, cooking oil (one brand). Cypermethrin Cooking oil (one brand). Quinalphos Fish, cooking oil (one brand). Aldrin Maize, fish, feed-mix, livestock feed, egg, human breast milk. Endosulphan Livestock concentrated feed, fish.

Source: Subba Rao (1999).

Table 7.3: Estimates of Daily Intake of DDT and HCH through Consumption of Milk Contaminated at Mean Level in Different States in India

State	*Intake (µg day⁻¹)*			
	Adult		*Child*	
	DDH	*HCH*	*DDH*	*HCH*
Andhra Pradesh	32.5	142.5	65.0	171.0
Bihar	10.0	44.5	12.0	53.4
Gujarat	22.5	17.7	27.0	21.3
Haryana	5.0	12.2	6.0	14.7
Himachal Pradesh	22.5	8.2	27.0	9.9
Karnatak	12.5	48.7	15.0	58.7
Kerala	7.5	19.7	9.0	23.7
Madhya Pradesh	10.0	12.7	12.0	15.3
Punjab	27.5	16.5	33.0	18.8
Uttar Pradesh	7.5	59.7	9.0	71.7
West Bengal	5.0	42.5	6.0	51.0

Source: Singh and Dhaliwal (1993).

The long term effect of these residues in human body include carcinogenicity, reduced fertility, increased cholesterol, high infant mortality and varied metabolic and genetic disorders (Subba Rao,

1999). Cases of blindness, cancer, diseases of liver and nervous system from pesticide poisoning have also been reported from cotton growing areas of Maharastra and Andhra Pradesh (Dhaliwal *et al.*, 2001).

Table 7.4: Monitoring of Pesticide Residues in Different Food Items

Food Items	*Number of Samples*		
	Analyzed	*Contained Residues*	*>MRL*
Vegetables	666	376	35
Fruits	317	118	1
Fennel	31	16	6
Cumin	35	18	6
Fenugreek	12	2	-
Chilly	45	27	-
Cotton	50	50	-
Fish	115	79	-
Butter	20	20	14

Source: Annual Progress Report (2004), All India Network Project on Pesticide Residues, (ICAR).

Effect of Pesticides in Soil

Large amount of pesticides reach the soil either directly by soil application or indirectly from the crops through the natural agencies. Excess amount of pesticides in soil may be harmful to soil microbes, germinating seeds and higher plants (Edwards, 1973). It has been reported that their presence in soil changes or alters fundamental soil properties like organic matter decomposition, nitrogen transformation, sulphur transformation, phosphate availability, trace element availability and soil enzyme activity. The fumigants and fungicides affect the soil microbial population and other beneficial organisms like earthworms etc. Their applications lead to increase in the NH_4–N, but nitrification is generally inhibited. Herbicides on the other hand, enhance the rate of nitrification. The hazards of pesticides in soil largely depend on their persistence. The longer they persist, the greater are the chances of contamination of soil. For persistence of organochlorine pesticides, it has been almost a dogma that the residues of these pesticides gradually increase as they pass from lower to higher trophic levels through the process of bio-magnification (Lodha *et al.*, 1989). The available information reveals that under tropical condition these pesticides are lost from soil into atmosphere in large amounts because of the hot and humid weather. It is evident that under moist conditions water molecules are preferentially absorbed and pesticide molecules being most non-ionic or anionic in character are set free and are lost by volatilization (Gupta, 1982). Novozhilov *et al.* (1974) observed the residues of Roger (dimethoate) in upper 30 cm layer of the soil for 40-60 days after application and reported that the rate of degradation increased with an increase in the moisture content in the soil from 5-29 per cent. The persistence of pesticides on soil to a large extent also depends on the soil types particularly soil pH and texture. Most pesticides are unstable under alkaline conditions and their persistence is comparatively longer in acidic and clay soils than in sandy and alkaline soils. The degradation of pesticides has also been found to undergo at faster rate under flooded condition (Rosen *et al.*, 1970). Under Indian condition, as there is a distinct rainy season it is likely that these pesticides quickly dissipate from the soil especially in rice growing fields. The importance of moisture in degradation of pesticides is attributed to the activity of anaerobic bacteria under flooded condition. Thus, lindane (gamma hexachloro cyclohexane), which is known to persist for several years in the aerobic upland soils, has been reported to undergo rapid decomposition in the tropical soils under flooding. Similarly many other pesticides also undergo rapid decomposition in soils under flooded conditions (Table 7.5).

In a recent trial conducted to study the adsorption the carbofuran in different soils of Orissa, it was observed that the adsorption of carbofuran was higher in black soils than in laterite soil. Leaching of carbofuran was reported to be in the order of laterite soil > alluvial soil > black soil. Maximum amount of carbofuran was retained within the upper 30 cm layer of soil columns in the black and alluvial soils whereas in black soils the movement occurred upto 15 cm. (Pal *et al.*, 1997). Degradation of pesticides in the soils also varies depending on the mobility of the concerned pesticides in the soil. Some pesticides are highly immobile whereas others show moderate to high mobility. Higher the immobility of pesticides, greater is their persistence in soil (Subba Rao, 1999). The compounds *viz.*, phorate, parathion, ethion, zineb, benomyl, heptachlor, endrin, aldrin, chlorodane, toxaphene and DDT etc. show high mobility. Many of these chemicals, however, are no more in use in agriculture. The commonly used herbicides, *viz.*, atrazine, alachlore, propachlore, simazine, propanil, diuron etc. have moderate to high immobility indicating greater persistence in soil. Besides, the persistence of a pesticide in soil also depends on its solubility in water. If the pesticide is very insoluble in water, it usually tends to stick to soil and also settle to the bottoms of the water bodies, making it less available to organisms. Water solubility and adsorption to soil are important in determining a pesticide's tendency to move through the soil profile with infiltrating water. The more strongly a pesticide sorbs to soil, the lower the tendency to move with infiltrating water. Soil properties are also important, as each soil has a characteristic ability to absorb pesticides. Soils high in clay and organic matter sorb pesticides better than sandy soils, which are low in organic matter. Soils that weakly adsorb pesticides and have a rapid infiltration rate are more sensitive to ground water pollution than soils that strongly adsorb pesticides and have a slow infiltration rate. Soil sorption and infiltration rate also determine pesticide loss in runoff. Soils with slow infiltration rate may be more prone to runoff, as more water will remain on the surface. Pesticides sorbed to soil will not be lost to runoff. However, if runoff results in soil erosion, pesticides sorbed to surface soil will also move with runoff. The potential for pesticide movement towards ground water or in runoff largely depends on pesticide properties, application factors as well as soil conditions. Rainfall, irrigation practices and evapotranspiration also significantly influence the potential for pesticide movement in water.

The degradation of pesticides is further accelerated in soil rich in organic matter and microbial biodiversity. On the other hand, in the soils, which are poor in organic matter the chemicals are bound to accumulate for longer period in soils without adequate microbial activities. The microbes present in soils significantly contribute to degradation of different chemical compounds through various metabolic activities *viz.* hydrolysis, oxidation, reduction and dechlorination etc. The bacteria predominantly involved in the degradation process are *Pseudomonas, Escherichia, Achromobacter, Arthrobacter, Bacillus, Clostridium, Corynobacterium, Flavobacterium* and *Micrococcus* etc. Among the actinomycetes, *Streptomyces* and *Thermoactinomyces* and among the fungi, *Aspergillus, Penicillium, Rhizopus, Trichoderma* and yeast species have been known to bring about degradation of pesticide to a large extent whether it is an organochloride, organophosphate, carbamate or any other chemical compound.

Table 7.5: Stability of Certain Pesticides in the Flooded Soils

Pesticides	*Period of Persistence (Days)*
Chlorinated hydrocarbons	
Alpha -HCH	30-90
Beta -HCH	30-90
Delta -HCH	30-90
Gamma -HCH	30-90
DDT	30-90
Methoxychlor	30-60
Heptachlor	30-90
Endrin	55
Organophosphates	
Diazinon	70
Parathion	20-60

* In non-flooded soils, chlorinated hydrocarbon insecticides included in this group persist for several years and organophosphates for six months, (Gupta, 1982).

Effect of Pesticides in Water

Pesticides drained out from crop fields often cause pollution of various water sources like rivers, ponds and lakes etc. which may prove lethal to the fish and other aquatic animals besides they being a threat to the human life. Pesticides enter surface and ground water primarily as runoff from crops and are most prevalent in agricultural areas. Pesticides, which are applied directly to the soil, may be washed off the soil into nearby surface water bodies or may percolate through the soil to lower soil layers and ground water. The potential of a pesticide to persist and contaminate water largely depends on its solubility in water. If a pesticide is very soluble in water, it serves as a potential groundwater contaminant through leaching. In addition, the water-soluble pesticide is more likely to get mixed in the surface water where it can have adverse effects on fish and other organisms. A number fish samples collected and analyzed in different parts of the country have shown residues of DDT, HCH, aldrin, heptachlore and dimethoate etc. (Dhaliwal *et al.*, 1994, 2001). The chlorinated hydrocarbons have shown greatest tendency for their biological concentrations. Traces of these pesticides in water may accumulate progressively in fishes and attain a concentration in their body tissues much higher than that in the surrounding water. DDT has been reported to accumulate in fish to levels more than 10,000 times the concentration present in the surrounding water (Gupta, 1982). Similarly, the unicellular aquatic organisms like planktons can also pick up the pesticide residue, which can get accumulated in them at a concentration much higher than its original concentration in the water. As the planktons serve as the food for fishes, the residues of pesticides enter the food chain. They are ultimately passed to the humans.

Effect on Resistance of Pests to Pesticides

Development of resistance in the target pests to the pesticides is yet another grave consequence of irrational use of pesticides with serious environmental implication. Resistance is the development of an ability to tolerate a dose of an insecticide, which would prove lethal to the majority of individuals in a normal population of the same species. Over-reliance on pesticides and their indiscriminate use have given rise to development of resistance to insecticides in more than 500 species of insect pests till 1990 (Dhaliwal *et al.*, 2001) all over the globe and 17 of these have been found resistant to all the available group of insecticides including DDT, cyclodines, organophosphates, carbamates and pyrethroides. In Asian countries, the first incidence of development of pesticide resistance was reported from India in *Singhara* beetle, *Galerucella birmanica* (Jacoby) that developed resistance to DDT and HCH. Since then 14 other pests have been reported resistant to different insecticides in one or more countries (Meherotra *et al.*, 2000). Armes *et al.* (1996) reported that resistance of *Helicoverpa armigera* to some pyrethroids is ubiquitous in this pest in the Indian sub-continent. The resistance levels to cypermethrin ranged from 5 to 6000- fold and fenvalerate 16 to 3200- fold. Pyrethroid resistance levels were recorded highest in central and southern India, which are intense cotton and pulse growing areas with extensive application of pesticides. Resistance to organophosphates was low and it ranged from 2 to 59-fold to quinalphos and 0.4 to 3-fold to monocrotophos. Another polyphagous pest, *Spodoptera litura* (Fabricious) that feeds on a number of crops including vegetables, oilseeds, cotton and tobacco crops has also been found to develop resistance to chlorinated hydrocarbons, organophosphates, carbamates and synthetic pyrethroids in India only in a span of 10 years, *i.e.* from 1971 to 1980. The level of resistance of this pest ranged from 0.2 to 197-fold to cypermethrin, 8 to 121-fold to fenvalerate, 1 to 13-fold to endosulphan, 1 to 29-fold to quinalphos, 2 to 362-fold to monocrotophos and 0.7 to 19-fold to methomyl (Armes *et al.*, 1997). During the period from 1965 to 1995 development of resistance in this pest has been reported to the extent of 21-fold to pyrethrum, 15-fold to endosulphan and 21-fold to lindane (Meherotra *et al.*, 2000). Due to indiscriminate use of pesticides in cotton, white fly (*Bemisia tabaci*) has developed resistance to a number of commonly used insecticides over the past few years. This pest has been reported to change its status from incidental to primary pest in different crops. A strain of white fly of brinjal was reported to exhibit more than 300-fold resistance to cypermethrin due to heavy reliance on synthetic pyrethroids (Cahil *et al.*, 1994). Similarly cotton leaf hopper (*Amrasca biguttula biguttula*) has also shown signs of resistance. Fenvalerate (0.15 per cent), which is 30 times the recommended concentration of 0.05 per cent, provided only less than 20 per cent mortality of this pest.

Pest Resurgence

Pest resurgence implies an abnormal increase in pest population following insecticide application often far exceeding the economic injury level. Brown plant hopper (*Nilaparvata lugens*) was a minor pest of rice in India during 1950s, but excessive use of pesticides has elevated it to the status of a major pest now not only in India but in whole of Asia. As many as 27 pesticides have been reported to cause resurgence in this pest (Dhawan *et al.*, 2000). Similarly, the use of synthetic pyrethroids on cotton during the last two decades also has resulted in increasing incidence of several pests including white fly, aphid, red mite and mealy bugs. Overuse of synthetic pyrethroids has been observed to be the major factor responsible for serious outbreaks of white fly in cotton (Jayraj, 1987) and aphid in mustard (Dhaliwal, 2001) in India during past few years.

The multifarious harmful consequences of indiscriminate use of pesticides have already cropped up as a serious threat to the ecosystem. In view of the fact that an increased use of pesticide has been drastically endangering the environmental sustainability, it is the need of the hour to go for an integrated approach in pest management rather than solely depending on pesticides. The integrated pest management (IPM) makes use of an ideal combination of physical, chemical, biological, genetical and biotechnological methods to contain crop damage below a level causing economic injury to the crops with minimum ecological implication. It takes the utmost advantage of the natural mechanism of pest suppression. Modification of the crop environment to make it unfavourable for infestation of pests, use of resistant cultivars and transgenic, conservation of predators and parasitoids together with a judicious and need-based use of preferably safer chemicals are the major components of IPM. Development of resistant varieties through conventional breeding procedures has remained as the most effective tool to overcome pest menace for years. Recently, biotechnology has also started playing an important role in the pest management strategy by developing transgenic plants. They produce insecticidal or antifeedant proteins continuously in plants under field conditions (Meeusen *et al.*, 1989). Bt. (*Bacillus thuringiensis*)–endotoxin genes are most commonly encountered in conferring resistance to plants because of established safety record, rapid action and complete biodegradability. Bt. toxins are stomach poisons. They act by disrupting the membrane of the insect gut disturbing the ionic balance and thereby paralyzing the gut. Of all the biological agents that have been evaluated so far, as pest control products, *Bacillus thuringiensis* has been the most successful insect pathogen (Dhaliwal *et al.*, 2001). Bt. Genes have now been produced in a large number of crops of economic importance like cotton, eggplant, maize, potato, rice, sugarcane, soybean, apple, chickpea and many other crop species. Many field trials of crops expressing these modified Bt. Genes have been carried out with impressive results. Some examples are *Cry1A* cotton against *Helicoverpa zea* and *Pectinophora gossypiella* (Saunders), *Cry3A* potato against *Leptinotarsa decemlineata* (Say) and *Cry1A* maize against *Ostrinis nubilalis* (Hubner). Bio-control is another integral component in ecofriendly pest management strategy. The pest population is kept under suppression by employing natural enemies like predators, parasitoids and antagonistic pathogens like viruses, bacteria and fungi. Biological control is a process in which population of one species lowers the population of another species by mechanism, *viz.* predation, parasitism, pathogenesis and competition. Success in biological control largely depends on its efficacy in the management of the pest population and complete success is achieved when the population of the target past is maintained below those causing economic injury to the crops in an extensive area. It eventually reduces the necessity of pesticide use and its load on the environment. Practical methods of conservation, multiplication and utilization of such living agents need to be further expanded for their wider applicability and acceptance. A judicious integration of different techniques into a unified pest management strategy is probably the only option left with us to ensure crop security from pest attack while maintaining the ecological balance at a reasonably safe level.

References

Armes N J, Jadava, D R and De Souza, K.R (1996) A survey of insecticide resistance in *Helicoverpa armigera* in Indian subcontinent. *Bull. Ent. Res.86:* 499-514.

Awasthi, M D (1987) Pesticide residues in horticultural crops (*Tech. Bull.*), IIHR, Bangalore.

Cahil, M, Johnston, D, Gorman, K and Denholm, I (1994) Insecticide resistance in *Bemisia tabaci* from Pakisthan. In: *Brighton Crop Protection Conference, Pests and Diseases.* Vol. 1. British Crop Protection Council, Farnham, UK, pp. 431-436.

Dhaliwal, G.S, Arora, R, Dhawan, A K and Singh, B (2000b) Intensive agriculture and pest problems: a case study of Punjab. *Indian J. Ecol.* 27(2): 109-130.

Dhaliwal, G S and Arora, R (2001) *Integrated Pest Management.* Kalyani Publishers, New Delhi.

Dhawan, A K, Dhaliwal, G S and Chelliah, S (2000) Insecticide–induced resurgence of insect pests in crop plants. In: *Pesticides and Environment.* Ed. Dhaliwal, G.S. and Singh, B. Commonwealth Publishers, New Delhi, India, pp.86-127.

Edwards, C A (1973) Pesticide residues in soil and water. In: *Environmental Pollution by Pesticides*, Plenumpress, New York. pp. 429-432.

Gupta, D K (1982) *Pesticides in Indian Environmrnt.* Interprint Publishers, New Delhi. ICAR (2004) All India Network Project on Pesticide Residues, Annual Progress Report.

Jayraj, S (1987) *Resurgence of Sucking Pests.* Tamil Nadu Agril. University, Coimbatore, India.

Kalra, R L, Singh, B and Battu, R S (1983) Studies on pesticide residues and monitoring of pesticidal pollution. Final Tech. Report, PL-480 Project, PAU, Ludhiana.

Kathpal, T S and Kumari, B (2000) Pesticide contamination of non-fatty food commodities. In: *Pesticides and Environment.* Ed. Dhaliwal, G.S. and Singh, B. Commonwealth Publishers, New Delhi, India, pp. 191-216.

Lodha, R M and Saxena, R C (1989) *Pesticides and Environmental Pollution*, Himansu Publications, Udaipur, India.

Mehrotra, K and and Phokela, A (2000) Insecticide resistance in insect pests: Current status and future strategies. In: *Pesticides and Environment.* Ed. Dhaliwal, G.S. and Singh, B. Commonwealth Publishers, New Delhi, India, pp. 39-85.

Mishra, B B and Mukherji, A K (2005) Sustainable Approaches in Agriculture- a Review. *Agril. Rev*, 26(3): 203-209.

Novozhilov, K V, Volkova, V A and Rozova, V N (1974) Dynamics of the degradation of dimethoate in plants and soil. *Khimiya v sel'skom Khoz.* 12 (3): 39-41 *Fide Soils Fertil.* 38(11): 397, Abstr. 5018; (1975).

Pal A K, Pattnaik, M R and Senapati, H K (1997) Movement, distribution and leaching of carbofuran in different soil columns. *Pesticide Res. J.* 9 (2): 165-168.

Rosen, J D, Siewierski, M and Winnett, G (1970) FMn-sensitized photolysis of chloroanilines. *J. Agric. Food. Chem.* 18: 494-96.

Singh, B and Dhaliwal, G S (1993) In: *Pesticides: Their Ecological Impact in Developing Countries.* Ed. Dhaliwal, G.S. and Singh, B. Commonwealth Publishers, New Delhi, pp. 131-161.

Singh, J and Dhaliwal, G S (1994) Insect pest management in rice: a perspective, In: *Trends in Agril. Insect Pest Management.* Ed. Dhaliwal, G.S. and Arora, R. Commonwealth Publishers, New Delhi, India, pp. 56-112.

Subba Rao, I V (1999) Soil and Environmental Pollution- A Threat to Sustainable Agriculture. The 26th Dr. R. V. Thamne Memorial Lecture, TNAU, Coimbatore, India.

Chapter 8

Biodiversity Utilization and Conservation in Hirakud Reservoir and Mahanadi River System: Freshwater Prawn Scenario

M.M. Prasad *, J.K. Bandyopadhyay and Prem Kumar [1]

Central Institute of Fisheries Technology, Burla Research Centre
Burla – 768 017, Sambalpur, Orissa
[1]Experimental Fish Farm and Field Center, NRC CWF,
Collecterate Road, Champawat, Uttaranchal – 262 523

ABSTRACT

The river Mahanadi and its tributaries as well as Hirakud Reservoir are replete with natural fauna. The freshwater prawn species, in these water resources are; *Macrobrachium malcolmsonii* (H. Milne Edwards), *Macrobrachium idae* (Heller), *Macrobrachium rude* (Heller), *Macrobrachium lamarrei* (H. Milne Edwards), *Macrobrachium lamarrei lamarrei* (H. Milne Edwards) *Macrobrachium dayanus* (Henderson), *Macrobrachium lendersonii*, *Caridina nilotica* (de Man) and *Caridina webari* (de Man). Among the nine identified species of prawn in Mahanadi River, *Macrobrachium malcolmsonii* is very much important due to its

* Corresponding Author: Ph: (O) 0663-2430503, (R) 0663-2432 619, (M) 9437736220, Fax: 0663-2430419;
E-mail: prasadmothadaka@yahoo.com, prasadmm@hotmail.com.

size and also fetches very high price. Large sized prawns are sold in local retail fish markets, on the riverbanks or on roadside at a higher price. While small sized prawns are disposed in the local markets at cheaper price or preserved by drying and smoking. In this paper the utilization of these prawn species for sustainable development and conservation of the species employing eco-friendly fishing gear are discussed.

Keywords: *Biodiversity, Conservation, Hirakud Reservoir, Mahanadi River, Prawn, Utilization.*

Introduction

The river Mahanadi emerges from the Sihawa hills to the extreme southwest of Raipur district of Chhattisgarh State drains though an area of 1,41,600 km^2, with a length of 857 km. Rivers such as Sheonanth, Oang, Tel and Ib are the tributaries of Mahanadi River System of which Ib is the largest one. These rivers are rain-fed which leads to their water flow during the monsoon between June and October. Hirakud the largest Reservoir on Mahanadi River System was taken up on the theme of Tennessee Valley project of America to control the devastating flood in Orissa due to Mahanadi River, power generation and irrigation. Total area of the reservoir quoted by Jhingran (1983) and Raghavachari and Rao (1984) is 74,592 hectares and 71,400 hectares respectively. Geographically, Hirakud is situated in-between 20°31′N, 85°10′ E latitude and 21°45′ N, 83°E longitude.

Freshwater Prawns in Mahanadi River System

The riverbed of Mahanadi River is rocky with creeks and crevices that provide natural habitat for freshwater crustaceans especially prawn. Behra and Rashid (2003) reported the presence of different freshwater prawn species, in the river Mahanadi, its tributaries and Hirakud Reservoir (Table 8.1). The same authors also reported the meager occurrence of *Macrobrachium rosenbergii*. The species has well established in the river Mahanadi. Harvesting of *Macrobrachium malcomsonii*, the pink gold of river has become a prime interest in resident fishermen on the banks of Mahanadi River since it fetches handsome amount of money.

Table 8.1: Occurrence of Prawns in Mahanadi River

Sl.No.	*Names of Prawn Species in Mahanadi River*
1.	*Macrobrachium malcomsonii* (H. Milne Edwards)
2.	*Macrobrachium idela* (Heller)
3.	*Macrobrachium rude* (Heller)
4.	*Macrobrachium lamarrei* (H. Milne Edwards)
5.	*Macrobrachium lamarrei lamarrei* (H. Milne Edwards)
6.	*Macrobrachium dayanus* (Henderson)
7.	*Macrobrachium landersonii*
8.	*Caridina nilotica* (de Man)
9.	*Caridina webari* (de Man)

Harvesting of Prawns

The crafts and gear employed in harvesting the prawns from the Mahanadi River and Hirakud Reservoir are not properly assessed or documented. The observation shows that fishermen catch prawn by Grasping (Hand picking), Three barbed pronged spear (Tenta and Barsha), Monoline (Daun), Pole and Line without hook (Garilathi), Trap (Benda), Bag net (Dhar jal and Chingdi jal), Falling Gear (Pheka, Thapa and Runi) and Gill net (Phansi jal). Crafts and Gears employed for the prawn fishing and need for increasing Catch Per Unit Efforts (CPUE) have been discussed in this chapter.

Post-harvest Utilization of Prawns

Smoke curing of freshwater prawn is an age-old practice prevalent in Western Orissa. This is one of the important sources of animal protein available to economically underprivileged people, specially the one residing in hinterlands. Among the nine identified species of prawn in Mahanadi River, *Macrobrachium malcolmsonii* is very much important due to its size and also fetches very high price. Large sized prawns are sold local retail fish markets, on the banks of river or on road side in fresh or in iced condition at a price of 140 to 160 rupees per kg. While small sized prawns are either sold in the local market at cheaper price or preserved by drying and smoking (Prem kumar and Meena Kumari, 2003). The products in general are disposed of with in a week.

The prawns are harvested from Mahanadi River System (MRS) and Hirakud Reservoir and are smoke cured in domestic kilns. In tropical countries, annual losses of cured fish due to bacterial spoilage amount to 2 to 3 million tons (Clucas and Ward, 1996). To under take studies on improvement of quality and shelf life it is essential to know the different quality aspects such as; physical, biochemical, microbiological and sensory characteristics of smoke cured prawns processed and sold at different locations. No data are available on these lines. Hence, in the present study commercial smoke cured prawn samples collected from different sources were subjected to quality analyses.

Materials and Methods

Fishing Methods

The information presented in this chapter was collected during survey on type of gears used for harvesting freshwater prawns in fishing villages near river Mahanadi such as Old Burla, Durgapali, Dhama and in Power Channel from Burla to Chiplima. Visits to the landing centres were made during different seasons of the year. The gears employed for prawn fishing were classified according to Brandit (1972).

Post-harvest Utilization

Quality Analyses of Commercial Sold Smoke Cured Prawns

Bacteriological Methods

The smoked prawn samples were collected from retail fish markets in self-locking new polythene bags. The bags were opened just before collection of sample to prevent external contamination to the product. Sampling for bacteriological analyses was carried out immediately after bringing it to the laboratory. 10 grams smoked prawn sample was weighed in aseptic condition and was cut into small pieces with sterile scissors under sterile conditions. The 10 grams cut pieces were aseptically triturated in sterile mortar with sterile pestle using part of 90 ml normal saline, which was later mixed with the rest.

Appropriate dilutions of the samples in sterile normal saline water were surface plated on plate count agar, violet bile salt glucose agar, KF streptococcal agar and Baird–Parker agar for the enumeration of mesophilic aerobes, faecal coliforms, faecal streptococci and staphylococci according to standard procedures (ICMSF, 1978; Mossel *et al*, 1978). Inoculated plates were incubated at 37°C for 24-96h, according to the type of enumerating bacteria. Colonies formed on the plates were counted and expressed as $\log_{10}$ cfu/g of the sample. Further, suspected colonies of *Staphylococcus aureus* were confirmed by coagulase positive test using rabbit plasma (Sanjeev and Surendran, 1996).

Throughout the study sterile normal saline (0.85 per cent NaCl) was employed as diluent. Chemicals and dehydrated media used in this study were of Qualigens (India) and Hi Media (India) make.

Physical and Biochemical Methods

The moisture, fats and αNH_2 (alpha amino nitrogen) in smoked prawn samples were estimated following AOAC (1995) methods. The total volatile nitrogen (TVN) in the fresh and cured fish samples was estimated by the method of Conway as described by Prasad *et al*. (1994).

External Quality of the Smoke Cured Prawns

The smoke cured prawns were screened for percentage of miscellaneous fish, percentage of extraneous material such as small parts fire wood material used for smoking, small molluscans, by sieving and insect infestation by manual methods using magnifying lenses. The miscellaneous fish (MF) or extraneous material (EF) in smoked prawns were assessed using following formula:

$$\frac{\text{Total weight of MF/EF}}{\text{Total weight of the specific sample}} \times 100 = \text{Percentage of MF/EF}$$

Sensory Evaluation

The smoke cured prawns were presented separately and evaluated for over all acceptability by a panel of 13 experienced panelists using a 10 point Hedonic scale. The same panel was employed throughout the study. The mean panel scores of each product were taken into consideration while assessing the quality. The quality of smoke cured prawn samples obtained from different sources, was assessed from the mean over all acceptability scores on source basis, assuming the Hedonic score '2' to be the limit, below which the product was not acceptable.

Results and Discussion

Fishing Methods of Freshwater Prawn in the River Mahanadi, Orissa

Fishing Craft

Fishing boats employed are locally known as 'dingis', which are plank-built canoes made of 'Rengal' (*Shorea robusta*), 'Bija' (*Pterocarpus marsupium*), 'Mohul' (*Madhuca longifolia*), 'Mango' (*Mangifera indica*), 'Gambaria' (*Gmelina arborea*), 'Semul' (*Bombax ceiba*) etc. Among these, the most commonly used wood is 'Rengal' or 'Sal', which is available locally in plenty. The dimensions of these canoes range from 5.5 m to 9.5 m in length and 0.43 m to 1.20 m in breadth. These canoes are of clinker type construction and are made locally by village carpenters. A coating of tar is generally applied on the boat as protection of the wood. The total cost of a 9 m canoe would be between Rs.6, 000 and Rs.10, 000 depending on the quality of wood employed. The life span of these boats is 5-7 years and their carrying capacity 2 to 4 fishermen with fishing gear and other accessories. These canoes are used in fishing and occasionally for transportation of fish catch to the landing centres. A few boats having slight modification in their design *viz.*, Caravel type came into operation later on with width considerably greater than that of the local canoes.

Fishing Gears

In harvesting of prawns different methods are employed. They included grasping by hand, wounding gears, Lines (Hook and line, Vertical line and Monoline), Pole and Line (with out hook), traps, bag nets, push nets, falling gears (cast nets) and gill nets. The method of operation and other information are discussed (Table 8.2).

Post-harvest Utilization

The commercial smoke cured prawn samples collected from different fish markets of Western Orissa, were assessed for bacteriological, physical, biochemical, external quality and sensory characteristics. Results obtained are presented separately under different sections of quality analyses.

Table 8.2: Gears and Crafts Operated for Harvesting Freshwater Prawns

Crafts and Gears	*Details of Craft and Gears Operated for Harvesting Prawns*
Craft operated	Fishing boats used for harvesting prawns are locally known as 'dingis', which are plank-built canoes made of 'Rengal' (*Shorea robusta*), 'Bija' (*Pterocarpus marsupium*), 'Mohul' (*Madhuca longifolia*), 'Mango' (*Mangifera indica*), 'Gambaria' (*Gmelina arborea*), 'Semul' (*Bombax ceiba*) etc.
Harvesting sans Gear: Hand picking	Fishermen catch the prawn by hand from the crevices in the stones. During night when shrimps comes out of crevices for food and migration fishermen use torch light to immobilize them and pick them by hand.
Wounding Gears	Spear with three barbed prongs locally known as *Barcha* or *Tenta*, which pierce in the body of the prawn. When gear is used in nights, torchlight is employed to search and immobilize the prawns.
Hook and line, Verticalline and Monoline	Hooks with small prawn and earthworm as baits attract prawns. Fishermen make a bait by using rice and spices, a commonly used preparation which is simple and easy to make, longer shelf life and more attractive to the prawn.
Pole and Line (with out hook)	This practice is seen during monsoon and early post monsoon season when water is turbid. A cotton line 2 to 3 meter lengths is tied with a bamboo stick called *Garilathi*. At the other end of line is pierced through an earthworm and a knot is put at the end. Fishing areas in the river are selected nears the boulders or where marginal weeds are submersed during rainy season. Prawn trying to eat worm gives a jerk and fishermen slowly start drawing out the line. Interestingly prawn doesn't leave the line. When it comes near the surface a handbag net is put down the prawn to make sure of its trapping. Generally 6 to 10 lines are operated at a time.
Traps	Traps are the major prawn fishing gears in Mahanadi. The traps are fixed near the creeks and crevices. Circular traps are made up of bamboo split locally known as *Benda*. Prepared bait is kept in side the trap through a conical entrance is made in the front side of the trap, which does not allow exit to the trapped animal. Gears are checked daily or alternately and are taken out through a window by opening it at the anterior side.
Bag nets	*Dhar Jal*, a conical shape of bag net of multifilament nylon is attached with a bamboo made cylindrical collection chamber at the posterior end. Iron ring is attached with the anterior portion or mouth, which provide support for setting and

Contd...

Table 8.2–Contd...

Crafts and Gears	*Details of Craft and Gears Operated for Harvesting Prawns*
	opening. The net is fixed against the water current with pages at both side of the ring. Prawn and fish are trapped in the collection chamber those have come through a conical entrance, which don't allow exit. This method is widely used in Old Burla and Dhama fishing villages.
Chingri Jal (push net)	This net is made of a rectangular piece of webbing (1.8 × 0.9 m) with a very small mesh size usually mosquito net. A thick nylon twine mounts the edge of the webbing and corners are attached with crisscrossed bamboo poles. This net is dragged along the shore in shallow waters for capturing small prawn (Khan *et al.*, 1992).
Cast nets	Falling gears like cast net (Pheka), Thapa and Runi are used in shallow zones of the river when water is turbid especially during monsoon season. These gears are used in the Ditches during post monsoon.
Gill nets	Gill nets, locally known, as *Phansi Jal* is most popular passive gear, which is used in rivers and reservoir. Although this net is not used exclusively for prawn fishing but if it is set near the hiding places of the prawn, they get entangled when they come out. In other method pieces of old fishing nets are put on the rocks for entangling the prawn.

Bacteriological Quality of the Smoked Products

Counts of different bacteria from all the smoked prawn samples obtained from different sources with averages are shown in Table 8.3. Fish or shellfish containing 8 log cycles per gram are unfit for human consumptions (Almas, 1981) and the values in the present study are with in limits (Table 8.3). Fecal coliforms occurred in 3 out of 12 samples screened (25 per cent). Occurrence of fecal streptococci is seen in 4 out of 12 samples (33.33 per cent). Their occurrence in more numbers and from more sources than faecal coliforms suggest that they are better indicators of hygiene conditions prevailing in post harvest handling and smoke curing places. Prasad and Seenayya (1998) also made similar observations. The mesophilic aerobic bacterial count of marine prawn caught from estuarine waters is at 3 log cycles, and the product is free from faecal coliforms and faecal streptococci (Table 8.3).

Table 8.3: Bacteriological Analyses of Smoke Cured Prawn Samples

Prawn Samples	*Mesophilic Aerobic Bacteria*	*Faecal Coliforms*	*Faecal Staphylo-cocci*	*Total Staphylo-cocci*	*Coagulase Positive Staphylo-cocci*
(Counts expressed in log cycles per gram of the sample)					
1	5,15	5,90	< 1	6,72	< 1
2	4,85	3,30	< 1	4,64	< 1
3	6,00	< 1	5,98	6,68	< 1
4	5,04	< 1	< 1	6,62	< 1
5	5,78	< 1	< 1	4,30	< 1
6	4,72	< 1	< 1	4,60	< 1
7	4,28	< 1	4,64	4,20	< 1
8	4,53	< 1	< 1	5,86	< 1
9	4,92	< 1	5,11	5,75	< 1
10	5,95	< 1	< 1	6,11	4,36
11	3,90	2,78	5,26	6,72	4,78
12	5,08	< 1	< 1	5,71	4,00
Average	5.42	4.80	5.04	6.27	3.89
13	3,60	< 1	< 1	6,41	< 1

Data presented are average of triplicate determinations.

Wherever the occurrence is nil, the results are shown as <1 log cycle, taking into consideration the sensitivity of the sampling method and limitations in selectivity of the media for promoting specific bacteria.

* Marine prawn from estuarine sources.

Sanjeev and Surendran (1996) reported staphylococcal food poisoning due to consumption of fish and fishery products. The mean staphylococcal count in the present study is 6.27 log cycles and is slightly higher than the threshold level (6 log cycles per gram) and is potential health hazard (Bergdoll, 1979). The coagulase positive staphylococci are distributed in 25 per cent of the samples with an average of 3.89 log cycles per gram of the samples. Sporadic occurrence of staphylococci from a wide variety of environmental sources has been reported (Kloos *et al*, 1992).

Physical and Biochemical Quality of Smoke Cured Prawns

Moisture plays an important role in quality of smoke cured prawns. In the present study the moisture content of all the samples ranged from 8.43 to 14.95 per cent and the average is 9.59 per cent (Table 8.4). The fat content ranged from 3.88 to 7.96 per cent with an average of 5.35 per cent, total volatile nitrogen 28.66 to 47.35 mg per cent with an average of 36.03 mg per cent and alpha amino nitrogen from 233.24 to 493.14 mg per cent with an average of 394.57 mg per cent. High total volatile nitrogen values were reported to correlate with high bacterial activity (Vanderzant *et al*, 1978). However, in the present study the level of TVN in the samples is far below the levels to indicate damage to the product. The moisture content of marine prawns (sample no.13) is similar to the average of other freshwater prawns. In the same marine prawns the fat and alpha amino nitrogen contents are less, and TVN is more in comparison to freshwater prawns (Table 8.4).

Table 8.4: Physical and Biochemical Quality of Smoke Cured Prawn Samples

Prawn Samples	*Moisture (%)*	*Fat (%)*	*Total Volatile Nitrogen (mg%)*	*Alpha Amino Nitrogen (mg%)*
1	14.95	5.91	47.35	233.24
2	8.43	5.48	35.51	493.14
3	8.79	4.30	34.28	463.17
4	12.67	5.07	33.64	333.20
5	7.84	5.23	34.27	446.49
6	8.07	7.34	39.87	446.49
7	7.86	7.96	47.35	473.14
8	9.01	4.43	28.66	233.24
9	11.30	6.15	43.61	359.86
10	8.88	3.88	29.90	379.85
11	8.83	4.03	29.28	466.48
12	8.50	4.44	28.66	406.50
Average	9.59	5.35	36.03	394.57
Standard Deviation	2.22	1.30	6.96	89.72
13*	9.52	1.81	93.45	239.90

* Marine prawn from estuarine waters.

Data presented are average of triplicate determinations.

External Quality of the Smoked Prawns

The results of external quality of product obtained are depicted in Table 8.5. Among 12 freshwater prawns samples screened 4 contained miscellaneous fish ranging from 0.47 to 3.44 per cent in the samples. The miscellaneous fish identified included *Xenontodon cancilla* (Gourchana-Oriya name), *Pama pama* (Patharmundi- Oriya name), *Rohitee cotio* (Chilanti-Oriya name), *Ambasis nama* (Patponia-Oriya name) and Puntius sp. Half of the samples contained extraneous material ranging from 0.14 to 3.13 per cent in the samples. The composition of extraneous material is large mud particles, small sticks, snails and burnt particles of algae. Insects were seen in one sample. However, no visible damage is observed. The marine prawn sample obtained from estuarine sources is free from miscellaneous fish, extraneous material and insect infestation.

Table 8.5: External Quality of Smoke Cured Prawn Samples

Prawn Samples	*Percentage of Miscellaneous Fish*	*Percentage of Extraneous Material*	*Insect Infestation*
1	Nil	2.74	-ve
2	Nil	Nil	-ve
3	Nil	Nil	-ve
4	Nil	Nil	-ve
5	3.44	3.13	-ve
6	Nil	Nil	-ve
7	0.47	2.49	-ve
8	Nil	Nil	-ve
9	Nil	Nil	-ve
10	4.05	0.14	+ve
11	1.32	1.52	-ve
12	1.32	1.56	-ve
13*	Nil	Nil	-ve

* Marine prawn from estuarine waters.
Data presented are average of triplicate determinations.

The small, unwanted fish that enter net during the harvesting of prawns need to be sorted out before starting the smoking process.

The percentage of other small fish in the product indicates inadequate care taken in post harvest handling. The percentage of extraneous material indicates that care taken was insufficient during the smoking process and the level of insect infestation indicates the improper storage methods after product preparation, but prior to marketing. The results indicate the need for improvement in all quarters of post harvest handling, till the product reaches the consumer. This is possible through training cum demonstration programs.

Sensory Characteristics

The results of sensory evaluation of freshwater smoked prawns and marine prawn sample are shown in Table 8.6. The over all mean score of thirteen panelists of the products show the range from 5.38 to 7.07 and majority of the samples scored at 6 level (75 per cent) and same pattern is also seen with marine prawn.

Table 8.6: Sensory Evaluation of Smoke Cured Prawn Samples

Prawn Samples		*Sensory Evaluation Scores*				
		Appearance	*Odour*	*Texture*	*Taste*	*Average*
1	Average[1]	6.50	6.42	6.31	6.58	6.45
	STD[2]	0.91	1.26	0.75	1.26	
2	Average[1]	6.98	7.03	6.65	6.73	6.85
	STD[2]	0.75	1.01	0.85	1.36	
3	Average[1]	5.85	6.42	5.92	6	6.05
	STD[2]	1.20	1.15	0.89	1.35	
4	Average[1]	6.23	6.08	6.04	6.87	6.31
	STD[2]	1.18	0.93	1.05	1.27	
5	Average[1]	6.42	6.27	6.19	6.58	6.37
	STD[2]	1.04	0.97	0.69	0.70	
6	Average[1]	7.15	6.88	6.19	7.38	6.9
	STD[2]	0.97	1.16	1.14	1.25	
7	Average[1]	6.92	7.07	6.88	7.42	7.07
	STD[2]	0.93	1.12	0.81	0.95	
8	Average[1]	6.81	6.31	6.46	6.62	6.55
	STD[2]	1.25	1.44	0.0366	1.28	

Contd...

Table 8.6–Contd...

Prawn Samples		Sensory Evaluation Scores				
		Appearance	*Odour*	*Texture*	*Taste*	*Average*
9	Average[1]	6.46	6.27	6.58	6.65	6.49
	STD[2]	1.03	1.48	0.81	1.07	
10	Average[1]	5.78	6.23	6.11	6.15	6.07
	STD[2]	1.29	1.15	0.77	0.88	
11	Average[1]	5.88	5.35	5.77	6	5.75
	STD[2]	0.82	1.44	1.01	1.41	
12	Average[1]	5.27	5	5.46	5.77	5.38
	STD[2]	1.17	1.31	1.39	1.42	
13*	Average[1]	6.88	6.15	6.58	5.96	6.39
	STD[2]	1.23	1.72	1.54	1.71	

The data of 1 and 2 represents the average and standard deviation of the sensory evaluation of 13 panelists who participated throughout the study.

* Marine prawn from estuarine waters. Results re-average of triplicate determinations.

Conclusions

Use of fishing gears of uncontrolled mesh size and fishing activities during breeding season has led to overexploitation of the resources. Greed for more catches has today resulted in reduction of mesh size of fishing gears to such an extent that even mosquito nets are used for fishing in shallow waters (Kumar *et al.*, 2002). The catches of mosquito net are mainly weed fishes and juveniles of commercially important fishes. This has resulted in hazard to the biodiversity of India's largest lacustrine ecosystem, which needs immediate repair by imposing ban on destructive gears and restocking of the reservoir.

The present study indicates the need for improvement in post harvest handling of the product to up grade over all quality, especially the hygiene aspects in view of occurrence of staphylococci in high numbers to an extent of hazardous to human health. In the present practice of smoke curing of fish significant amount of firewood is collected from the forest that is resulting in deforestation. After years of research the Burla Research centre of Central Institute of Fisheries Technology has come out with different models of Community Fish Smoking Kilns what could be termed as Green kilns that supports the conservation of biodiversity of flora of Hirakud Reservoir area. In the Green kilns the firewood in very less quantities is employed in the initial stages of smoking and rest is replaced by paddy husk or saw dust to generate smoke.

Conservation of biodiversity is one of today's thrust areas and to achieve this coordination between research, academic institutions, governmental (state fisheries department) and non-governmental is highly essential to make efforts towards awareness programme and dissemination of advanced technologies. Updating of the baseline data on the reservoir and riverine fishing techniques is part of the effort to evaluate the present status catering the future needs. The Burla Research Centre of CIFT, taking into prime consideration, the conservation of biodiversity and awareness drive in fishermen community, has taken up programs such as; Village adoption, conducting of training cum demonstration programs *viz.*, 1. Hygienic preparation of smoked cured fish. 2. Preparation of value added products. 3. Use of eco-friendly gears for sustainable fisheries etc., in remote fishing villages adjoining Hirakud Reservoir.

Acknowledgement

Authors are thankful to Dr. Devadasan, the Director, CIFT, Cochin, for his permission to presentation and publication of this paper in National Symposium on "Environmental Biotechnology and Biodiversity Conservation".

References

Almas, A K (1981) Chemistry and microbiology of Fish and Fish Processing. Dept. of Biochemistry, Norwegian Institute of Technology, University of Trondheim, Norway.

AOAC (1995) *Official Methods of Analysis*, 16th Edn. Association of Official Analytical Chemists, Washington, D.C, USA.

Behra, M K and Rashid, A (2003) On Biology of River Prawn (*Macrobrachium malcomsonii*), its Capture Fishing and Culture in Western Orissa. Fishing Chimes. 22 (10 and 11) 60-61 and 90.

Bergdoll, M S (1979) In: Reiman, H, Bryan, F.L (Eds.) Foodborne infections and intoxications. Academic Press. New York, USA, pp 444.

Brandit, A V (1972) Fish Catching Methods of the World. Revised and Enlarged Edition.

Clucas, I J and Ward, A R (1996) *Post harvest fisheries development: A guide to handling, preservation, processing and quality*. Chatham, Maritime, Kent ME 4 4TB. United Kingdom.

Conway, E J (1947) *Microdiffusion analysis and volumetric error*. Crossby. Lock Wood and sons. London. U.K.

Jhingran, V G (1983) Fish and Fisheries of India. Hindustan Publishing Corpn. Delhi. 170-199.

ICMSF (1978) Micro-organisms in foods. 2nd edn. International Commission on Microbiological Specifications for Foods. University of Toronto Press, Toronto, Canada.

Khan, A A, Dawson, P, Rao, J S R and Varghese, M D (1992) Fishing in Impounded Waters. A Case Study of Hirakud Reservoir-Orrisa. CIFT Publication.

Kloos, W E, Schleifer, H K, Gotz, F (1992) The genus Staphylococci. In: Ballows A, Truper, H.G, Dworkin, M, Harder, W, Schleifer, K.H. (Eds.) The Prokaryotes: A Hand Book on the Biology of Bacteria; Ecophysiology, Isolation, Identification and Applications. Springer-Verlag. New York, pp 1369-1420.

Kumar, P, B Meenakumari and J K Bandyopadhyay (2002) Biodiversity Conservation of Hirakud Reservoir. J. Natcon. 14(1): 147-150.

Mossel, D A A and Bijker P G H, Eelderink, I (1978) Streptococci of Lancefield groups A, B and D and those of *Buccal* origin in foods; Their public health significance, monitoring and control. In: Skinner, F.A and Quennel, L.B (Eds.) *Streptococci*. Academic Press. London, pp. 315-333.

Prasad, M M, Panduranga Rao CC and Gupta SS (1994) "*Chemical and microbial quality of dry fish from Kakinada area.*" Fishery Technol. 32(1): 75-78.

Prasad, M M, Seenayya, G (1998) Major Microbial Contamination Points (MMP) in Fish Curing Environments of India's Andhra Coast. J. Food Sci. Technol, 35(5),458-460.

Prem Kumar and Meenakumari, B (2003) Methods of Prawn Harvesting from The River Mahanadi, Orissa.

Raghavachari, M and Rao, S.S (1984) Inland Fisheries Resources in India. Concept Publishing Company New Delhi. 213-250.

Sanjeev, S, Surendran, P K (1996) Fate of enteropathogenic Staphylococci in fish subjected to curing. Fish Technol. 33(1), 66-68.

Vanderzant, C, Cobb, B F, Thompson, C A Jr (1973) Microflora, chemical characteristics and shelf life of four species of pond reared shrimps. *J. Milk Food Technol*. 35, 443-446.

Chapter 9

Mangroves of Orissa Coast with Special Reference to Mahanadi Delta

R.K. Nayak [1] and B.P. Choudhury [2]

[1]*Department of Botany, G.M. College, Sambalpur – 768 004, India*

[2]*Department of Botany, Utkal University, Bhubaneswar – 751004, India*

ABSTRACT

Mangroves of Mahanadi delta in the Orissa Coast was much rich in remote past. In course of time, this vegetation has been depleted at an alarming rate due to the operation of various biotic factors such as, the establishment of Paradeep Port, Paradeep Phosphates Ltd., piscicultures, settlement of immigrants etc. As a result, the present vegetation is extant in most denuded condition, which needs an urgent attention for conservation.

Keywords: *Conservation, Mahanadi delta, Mangroves, Orissa coast.*

Introduction

Mangroves are peculiar groups of salt tolerant plants constituting important vegetation in different estuarine belts of the world as these are important source of timber, fire wood, tanin etc. Besides they play an important role in the coastal environment by protecting the shoreline and checking the intensity of tropical cyclones. Mangroves of India constitute 7 per cent of the world mangroves covering an area of 6740 sq. kms. (Anonymous, 1987). There is significant distribution of mangroves and their associates in different parts of the Orissa coast such as Mahanadi delta, Bhitarkanika, Dhamara estuary etc. Among them Mahanadi delta harbours an important manghal formation in the Indian sub-continent. It lies between 19° 45′–20° 30′ N latitude and 85° 15′–86° 50′ E longitude covering an area of 13600 sq. kms. Of these, only 120 sq. kms. is covered with mangrove forests in differenet eustarine zones of Mahanadi delta. This area enjoys the tropical climate temperature of this region ranges from 21°-40°C. The annual rainfall is 141.15 cm. The soil is sandy in the fringes of the coast and alluvial silty along the mouth of the river Mahanadi.

From the past floristic studies, it is evident that the flora of this region including the mangroves were much rich in remote past. However, these have been degraded considerably due to the operation of various biotic factors which includes the inception of Paradeep Port, Paradeep Phosphates Ltd., piscicultures, settlement of immigrants etc. As a result, the present vegetation is extant in denuded condition. The present investigation gives a comprehensive idea about the distribution of various mangrove species in different locations of Mahanadi delta. This was in connection with a project entitled "Ecofloristic studies in the Mahanadi delta with special reference to mangroves and their associates" sponsored by the University Grants Commission, New Delhi.

Materials and Methods

Literatures pertaining to the past floristic studies in the Mahanadi delta have been consulted in order to have a comprehensive idea about the distribution of various mangrove species and their status in remote past. Herbarium specimens of various mangrove species deposited in the Central National Herbarium (CNH), Calcutta have been consulted to have a clear idea about the distribution of various mangrove species in remote past. Regular field trips have been conducted to different areas of the Mahanadi delta in different seasons of the year to have a clear idea about the present floristic composition of the mangroves and their associates extant in this region. Herbarium samples of different species have been prepared which are housed in the herbarium of Post Graduate Department of Botany, Utkal University, Bhubaneswar.

Results and Discussion

From the past floristic studies conducted by a number of workers such as Haines (1921-25), Mooney (1950), Rao and Banerjee (1982), Rao *et al.* (1970), Choudhury *et al.* (1991), Subudhi *et al.* (1992), Choudhury (1994), it is evident that the mangroves of Mahanadi delta was much rich in remote past. However in course of time, these have been degraded to much extent due to the operation of various biotic pressures as mentioned earlier.

During the present floristic studies 46 species of mangroves and their associates including one member of pteridophyta have been collected from different areas of Mahanadi delta (Nayak, 2002). It is also to be noted here that Mangroves of Bhitarkanika are better protected compared to that of Mahanadi delta as it has been declared as a wild life sanctuary by the Government of India. Up till now 67 species of mangroves and their associates have been reported from Bhitarkanika. A comparative list of distribution of mangrove species in Mahanadi delta and Bhitarkanika has been shown in Table 9.1. It is evident that the mangroves of Mahanadi delta has been degraded at an alarming rate due to the operation of various biotic factors cited earlier (Nayak and Choudhury, 2002). As a result, the present mangrove vegetation is extant in denuded condition in different areas such as Paradeep, Batighar, Hukitola Island, Jambu, Ramnagar, Jatadhar Muhan, and Devi estuary. Of these, the status of the mangroves of Paradeep is much alarming due to the clearing of mangrove forests for the construction of Paradeep Port and Paradeep Phosphates Ltd. The present mangrove elements are extant only in the defunct creeks of Ghanagolia and Atharabanki and Paradeep Port area. The dominant elements of the present vegetation of this area includes species like *Acanthus ilicifolius, Phoenix paludosa, Tamarix troupii, Avicennia officinalis, Rhizophora mucronata* etc.

The status of mangrove elements of Jambu island is equally alarming due to the acute anthropogenic pressure which has resulted due to the settlement of immigrants. The present mangrove vegetation is represented by few species of mangroves and their associates such as *Acanthus ilicifolius, Phoenix paludosa, Tamarix troupii, Hibiscus tiliaceus, Porteresia coarctata* etc.

Table 9.1: Comparison of Distribution of Mangrove Species and their Associates in Mahanadi Delta and Bhitarkanika

Sl.No	Name of the Species	Mahanadi Delta	Bhitarkanika
1.	*Acanthus ilicifolius* L.	+	+
2.	*Acrostichum aureum* L.	+	+
3.	*Aegialitis rotundifolia* Roxb	+	+
4.	*Aegiceras corniculatum* (L.) Blanco	+	+
5.	*Aglaia cucullata* (Roxb.) Pellegrin	–	+
6.	*Avicennia officinalis* L.	+	+
7.	*Avicennia alba* Bl.	+	+
8.	*Avicennia marina* (Forssk.) Vierh.	+	+
9.	*Brownlowia tersa* (L.) Kostern	+	+
10.	*Bruguiera cylindrica* (L.) Bl.	+	+
11.	*Bruguiera gymnorrhiza* (L.) Savigny	+	+
12.	*Bruguiera parviflora* (Roxb.) Wt and Arn. ex Griff.	–	+
13.	*Bruguiera sexangula* (Lour.) Poir.	–	+
14.	*Caesalpinia bonduc* (L.) Roxb.	+	+
15.	*Caesalpinia crista* L.	+	+
16.	*Cerbera manghas* L. *Syn. Cerbera odollam* Gaertn.	–	+
17.	*Ceriops decandra* (Griff.) Ding	+	+
18.	*Ceriops tagal* (Perr.) C.B. Rob.	+	+
19.	*Clerodendrum inerme* (L.) Gaertn.	+	+
20.	*Cynometra iripa* Kostel	+	+
21.	*Cynometra ramiflora* L.	–	+
22.	*Cyperus malaccensis* Lamk.	+	+
23.	*Dalbergia spinosa* Roxb.	+	+
24.	*Derris scandens* (Roxb.) Bentham.	+	+
25.	*Derris trifoliata* Lour	+	+
26.	*Excoecaria agallocha* L.	+	+
27.	*Sapium indicum* Arg. Willd.	–	+
28.	*Fimbristylis ferruginea* (L.) Vahl.	+	+
29.	*Finlaysonia obovata* Wall.	+	+

Contd...

Table 9.1–Contd...

Sl.No	Name of the Species	Mahanadi Delta	Bhitarkanika
30.	*Flagellaria indica* L.	–	+
31.	*Heritiera fomes* Buch.- Ham.	+	+
32.	*Heritiera littoralis* Dryand	–	+
33.	*Heliotropium curassavicum* L.	+	+
34.	*Hibiscus tiliaceus* L.	+	+
35.	*Instia bijuga* (Colb.) Kunth.	–	+
36.	*Ipomoea pes-caprae* (L.) R. Br.	+	+
37.	*Kandelia candel* (L.) Druce	+	+
38.	*Lumnitzera racemosa* Willd.	+	+
39.	*Merope angulata* (Kurz) Swingle	+	+
40.	*Mucuna gigantia* (Willd.) DC.	–	+
41.	*Myriostachya wightiana* (Nees ex Steud.) Hook. f.	+	+
42.	*Phoenix paludosa* Roxb.	+	+
43.	*Porteresia coarctata* (Roxb.) Tateoka	+	+
44.	*Rhizophora apiculata* Bl.	+	+
45.	*Rhizophora mucronata* Poir.	+	+
46.	*Rhizophora stylosa* Griff.	–	+
47.	*Sapium indicum* Arg. Willd.	–	+
48.	*Salacia prinoides* (Willd.) DC.	–	+
49.	*Salicornia brachiata* Roxb.	+	+
50.	*Salvadora persica* L.	+	+
51.	*Sarcolobus carinatus* Wall.	–	+
52.	*Sarcolobus globosus* Wall.	+	+
53.	*Schoenopletus littoralis* (Schrad.) Palla	+	+
54.	*Sesuvium portulacastrum* (L.) L.	+	+
55.	*Sonneratia alba* J. Smith	+	+
56.	*Sonneratia apetala* Buch-Ham.	+	+
57.	*Sonneratia caseolaris* (L.) Engl.	–	+
58.	*Sonneratia griffithii* Kurz.	–	+
59.	*Suaeda maritima* (L.) Dumort.	+	+

Contd...

Table 9.1–Contd...

Sl.No	Name of the Species	Mahanadi Delta	Bhitarkanika
60.	*Suaeda nudiflora* (Willd.) Moqe	–	+
61.	*Tamarix troupii* Hole	+	+
62.	*Tamarix ericoides* Rottl	-	+
63.	*Tamarix dioica* Roxb.	–	+
64.	*Thespesia populnea* (L.) Sol. ex Correa	+	+
65.	*Thespesia populneoides* (Roxb.) Kostel.	–	+
66.	*Tylophora tenuissima* (Roxb.) Wt. and Arn. ex Wt.	+	+
67.	*Xylocarpus granatum* Koeing	–	+
	Total Number of Species	***46***	***67***

The + sign indicates the presence of the species and–sign indicates the absence of the species in the particular area.

Mangroves of Batighar area is rather more gregarious along the creeks due to the public conciousness and the plantation programmes for the restoration of mangrove vegetation which has been initiated by some agencies like M.S. Swaminathan Research Foundation with the active participation of the local people. Some dominant element extant in this region includes *Rhizophora apiculata, Rhizophora mucronata, Kandelia candel, Sonneratia apetala* which are much gregarious.

Magroves of Hukitola island is better protected as it is subjected to less biotic pressure due to geographical isolation. Some important mangrove elements extant in this area include *Aegialitis rotundifolia, Lumnitzera racemosa, Rhizophora apiculata, Tamarix troupii, Thespesia populneoides.*

The mangrove vegetation in Kharnasi, Ramnagar is extant in denuded condition which is represented only by some mangrove associates such as *Acanthus ilicifolius, Porteresia coarctata, Tamarix troupii, Hibiscus tiliaceus* etc. Most of the lands in this area have been converted into paddy fields.

From the present investigation it has been observed that the mangrove vegetation of Mahanadi delta in general and that of the Paradeep in particular is extant in denuded condition. It is also to be noted here that 16 species of mangroves and their associates reported by earlier workers (Nayak and Choudhury, 2002) could not be traced during the present investigation (Table 9.2).

Table 9.2: List of Mangroves and their Associates Reported by Earlier Workers in Mahanadi Delta but could not be Traced During the Present Study

Sl.No	Name of the Species	Family
1.	*Aglaia cucullata* (Roxb.) Pellegrin	Meliaceae
2.	*Cerbera odollam* Gaertn.	Apocynaceae
3.	*Heritiera littoralis* Dryand ex Ait.	Sterculiaceae
4.	*Instia bijuga* (Collbr.) Kuntze.	Caesalpiniaceae
5.	*Mucuna gigantea* (Wlld.) DC.	Fabaceae
6.	*Rhizophora stylosa* Griff.	Rhizophoraceae
7.	*Salacia chinensis* L.	Hippocrateaceae
8.	*Sarcolobus carinatus* Wall.	Asclepiadaceae
9.	*Sonneratia caseolaris* (L.) Engl.	Sonneratiaceae
10.	*Sonneratia griffithii* Kurz.	Sonneratiaceae
11.	*Suaeda monoica* Forssk. ex Geml.	Chenopodiaceae
12.	*Tamarix ericoides* Rottl.	Tamaricaceae
13.	*Tamarix dioica* Roxb.	Tamaricaceae
14.	*Thespesia populneoides* (Roxb.) Kostel.	Malvaceae
15.	*Xylocarpus granatum* Koeing	Meliaceae
16.	*Xylocarpus mekongensis* (Prain) Pierre.	Meliaceae

This has created various environmental hazards like tropical cyclones, encroachment of sea towards land, floods etc. This is clearly evident from the devastations caused in the last Super cyclone of October '99 which was accompanied with high speed of wind and unusual rise of sea water from the Bay of Bengal that has washed away many lives of human beings, cattle and other animals. Large number of trees has been uprooted and severely damaged in the coastal regions of Orissa. It is to be noted here that the Bhitarkanika area was least affected in the Super cyclone due to thick mangrove vegetation.

In situ conservation of the present mangrove vegetation is an urgent need of the hour. Besides this, natural regeneration of some species of mangroves and their associates which are better adopted in this region such as *Rhizophora apiculata, Rhizophora mucronata, Kandelia candel, Avicennia officinalis, Avicennia alba, Sonneratia apetala, Excoecaria agallocha, Phoenix paludosa* etc. should be done on top priority. Mass propagation of some of these species should be done for extensive plantations in this region. The successful plantation of *Rhizophora apiculata* in the south of Ho Chi Minch City of Vietnam which exhibits luxuriant growth is a bright example of mangrove plantation in denuded regions. Public awareness should be generated through seminars, symposia, workshops, distribution of pamplets, posters etc.

State as well as the Central Government should take various measures for the conservation, regeneration as well as rehabilitation of the mangrove vegetation with the active participation of voluntary agencies and the local people. This will be very useful for the restoration of the past vegetation and maintainance of environmental equilibrium in this region.

Acknowledgements

Authors are thankful to the Professor and Head, Department of Botany, Utkal University, Bhubaneswar for providing laboratory facilities and to the authorities of Central National Herbarium for utilising the Herbarium facility. The first author is also thankful to the authorities of University Grants Commission for financial assistance in the form of minor research project.

References

Anonymous (1987) *Mangroves in India: Status report*. Government of India, Ministry of Env. and Forests, New Delhi.

Choudhury, B P (1994) A Comparative study of mangrove vegetation in Bhitarkanika and Paradeep area in the state of Orissa: 33-41 *in Forest, Wildlife, Environment*. Andhra University, Visakhapatnam (Ed. M.V. Subba Rao).

Choudhury, B P, Biswal, AK and Subudhi, H N (1991) Mangroves of Orissa and aspects of their conservation. *Rheedea* 1: (1 and 2) 62-67.

Haines, H H (1921-1925) *The Botany of Bihar and Orissa* (1-6), London.

Mooney, H F (1950) *Supplement to the Botany of Bihar and Orissa*. Catholic Press, Ranchi.

Nayak, R K (2002) *Ecofloristic studies in Mahanadi delta with special reference to Mangrove and their associates*. PhD. thesis, Utkal University, Bhubaneswar.

Nayak, R K and Choudhury, B P (2002) Status of mangroves of Mahanadi delta:Past Present and Prospects for future. *Asian Jr. of Microbiol. Biotech. Env. Sc.* 4(1): 93-97.

Nayak, R K and Choudhury, B P (2002) Present status and conservation strategies of mangroves in Mahanadi delta. Proc. National.Seminar on conservation of Eastern Ghats. Tirupati pp. 275-279.

Rao, T A and Banerjee, L K (1967) Some new records for Orissa state. *J. Bomb. Nat. Hist. Soc.,* 64: 583-584.

Rao, T A, Banerjee, L K and Mukherjee, A K (1970) Some interesting plant records from Orissa Coast. *J. Bomb. Nat. Hist. Soc.* 66. 659-660.

Rao, T A and Banerjee, L K (1982) Tidal Mangroves of the Mahanadi Delta, Utkal Coast, India. *Proc. Seminar. Res. Devt. and Env. in Eastern Ghats,* Waltair, pp 219-226.

Subudhi, H N, Choudhury, B P and Acharya, B C (1992) Potential Medicinal Plants from Mahanadi delta in the state of Orissa. *J. Econ. Tax. Bot.* 16: 479-487.

Chapter 10

Study of Algal Diversity from Rice Fields of Baripada, Mayurbhanj, Orissa

S. Bhakta, H. Dey and A.K. Bastia*

P.G. Department of Botany, North Orissa University, Baripada – 757003, Orissa

ABSTRACT

Algae constitute a dynamic component of soil, which aid and increase the availability of crop nutrients. A taxonomic survey of algal flora from rice fields of Baripada in the Mayurbhanj district, Orissa was carried out during January to June 2006. A total of twenty algal taxa belonging to eighteen genera were collected out of which six belonging to Cyanophyceae, eight to Chlorophyceae, five to Bacillariophyceae and one to Charophyceae. The predominant Cyanophyceae members were *Oscillatoria, Phormidium, Lyngbya, Anabaena, Nostoc* and *Cylindrospermum.* Both heterocystous and non-heterocystous forms were found and most of them occur in floating form. The common Chlorophyceae were *Scenedesmus, Zygnema, Spirogyra, Closterium, Cosmarium, Pediastrum* and *Oedogonium. Scenedesmus* is represented by two

* Corresponding Author: Reader in Botany, E-mail: bastiaaknou@yahoo.co.in.

distinct species. The common Diatoms identified were members of *Fragilaria, Pinnularia, Nitzschia* and *Cymbella. Chara connivence* is the single Charophyceae reported during the survey.

Keywords: *Algal diversity, Baripada, Rice fields.*

Introduction

Algae are simplest plants inhabiting almost all kind of habitats. They constitute a dynamic component of the soil, increase soil productivity and nutrient utilization efficiency of the plants. It is therefore essential to study the algal distribution in nature for their better exploitation. The information on algal distribution in Orissa is scanty and most of the work has been restricted on blue green algal form of rice fields (Adhikary, 1998; Mohanty and Padhi, 1982,1984; Parija *et al.*, 1969; Sahu *et al.*, 1996; Sahu, 2000). Baripada, the head quarter of Mayurbhanj district of Orissa state, is situated at the foot bed of Similipal Biosphere Reserve within North Central Plateau. The climatic condition of the district is subtropical in nature and is characterized as 'extreme' in hot summer, raises temperature upto 43°C and the temperature in winter falls to 8°C. The pH of the soil of the rice fields varies from 6.5 to 7.2. A perusal of the literature shows that algal diversity of this area of the state has not been studied so far. Hence, the present work was undertaken with a view to study the algal diversity from rice fields of Baripada and adjoining regions.

Materials and Methods

Algal samples were collected from rice fields of different localities of most of the Grampanchayats of Baripada during January to June 2006 at different stages of the Rabi crop. The samples were collected in clean sterilized collection bottles using forcep and needle and brought to the laboratory for identification. The approximate pH of water from various sites was measured with the help of universal pH paper simultaneously. Each sample was assigned with a collection number along with place and date of collection, fixed in 4 per cent formaldehyde. Temporary slides were made and examined under compound research microscope. Microphotograph of each specimen was taken soon using a Meiji Trinocular research microscope fitted with Nikon F10 Camera. The organisms were identified following standard literatures and available monographs of Desikachary (1959), Randhwa (1959), Gonzalves (1981), Anand (1998).

Systematic Enumeration

Cyanophyceae

***Oscillatoria chalybea* (Martins) Gomont (Plate 10.1, Figure 1) Desikachary, 1959, p.218, Pl.38, Figure 3.**

Thallus dark blue green; trichome straight, slightly constricted at the crosswalls, attenuated at the apex and somewhat bent, septa not granulated, end cell obtuse, not capitate, without calyptra.

Floating; Collection no. SC (6); Collection site: Khetrapatana rice field; pH 6.8; Date of collection: 11.04.2006.

***Phormidium purpurascens* (Kutz.) Gomont (Plate 10.1, Figure 2) Desikachary, 1959, p.262, Pl. 45, Figure 1.**

Thallus compact, leathery, purple to brownish violet; trichome entangled, not constricted at the cross walls, cells nearly quadrate or upto nearly two times longer than broad, cross walls marked by two granules on either side, end wall rounded, calyptra absent.

Attached to the soil; Collection no. SA (10); Collection site: Raghunathpur rice field; pH 7.0; Date of collection: 08.04. 2006.

***Lyngbya ceylanica* Wille (Plate 10.1, Figure 3) Desikachary, 1959, p. 299, Pl. 54, Figure 4.**

Thallus olive green, violet or red; filament straight, unconstricted at the cross walls, not attenuated at the ends, cross walls not granulated, cells quadrate, broader than long, calyptra absent.

Attached to the soil; Collection no. SA (10); Collection site: Raghunathpur rice field; pH 7.0; Date of collection: 08. 04. 2006.

***Anabaena constricta* Skuja (Plate 10.1, Figure 4) Desikachary, 1959, p.419, Pl.71, Figure 3.**

Cells cylindrical, both end truncated, blue green, trichome single, not forming a thallus, with cells distinctly placed from one another.

Free floating; Collection no. SC (1); Collection site: Khetrapatana rice field; pH 7.0; Date of collection: 11.04.2006.

Nostoc spongiaeforme **Agardh ex Flah. (Plate 10.1, Figure 5) Desikachary, 1959, p. 380, Pl.67, Figure 3.**

Thallus at first globose, gelatinous, later expanding, light blue green, filaments flexuous, loosely entangled; cells partly cylindrical, partly short barrel shaped; heterocysts sub spherical.

Free floating; Collection no. SA (8); Collection site: Manchabandha rice field; pH 6.8; Date of collection: 18.03 2006.

Cylindrospermum stagnale **F. variabilis Prasad, B.N. (Plate 10.1, Figure 6) Desikachary; 1959, p.363, Pl.64, Figure 1.**

Plant mass soft, dense, mucilaginous, light blue green, forming a mat; trichome single, heterocyst at end of filament, elliptical.

Fixed to soil; Collection no. SA (5); Collection site: Raikot Jharana rice field; pH 7.2; Date of collection: 02.03.2006.

Chlorophyceae

Pediastrum **sp. (Plate 10.1, Figure 7)**

The colony is stellate shaped. Eight cells are in the colony grouped together into a non-motile colony. The cells are arranged in a concentric ring around a central cell. The cells are polyhedral in shape.

Floating; Collection no. SA (6); Collection site: Palbani rice field; pH 6.8; Date of collection: 03.04.2006.

Scenedesmus obliquus **(Turp.) Kutz. (Plate 10.1, Figure 8)**

Four celled colonies, cells arranged in linear series, fusiform cells.

Submerged floating; Collection no. SA (6); Collection site: Palbani rice field; pH 6.8; Date of collection: 03.04.2006.

Oedogonium excavatum **Jao Figure 5 (Plate 10.1, Figure 9) Gonzalves, 1981, p.196, Figure 9.66A.**

Macrandrous, monoecious; vegetative cells cylindric, elongated cells, oogonium single, sub ellipsoid, oospore nearly filling the oogonium; wall three layered.

Fixed floating; Collection no. SA (8); Collection site: Betna rice field; pH 6.6; Date of collection: 02.03.2006.

***Scenedesmus* sp. (Plate 10.2, Figure 1)**

Colonies of 8 cells, having minute spines through out the wall of cells. Each end cell containing two spines, one is little small in size than other one.

Floating; Collection no. SA (6); Collection site; Palbani rice field; pH 6.8; Date of collection: 03.04. 2006.

***Zygnema mirandum* TAFT 1946. Trans. Amer. Micros. Soc. 65, p.24. (Plate 10.2, Figure 2)**
Randhwa 1959, p.242, Figure 196.

Filaments are unbranched with short cylindrical cells. Vegetative cells are longer than the broad with two stellate chloroplasts with a prominent central pyrenoid, one on either side of a centrally situated nucleus.

Floating; Collection no. S7; Collection site: Debendrapur rice fields; pH 7.0; Date of collection: 20.03.2006.

***Spirogyra schmidtii* W. and G.S. WEST 1902. Bot.Tiddsskr. 24, p.161, Figs. 43-45. (Plate 10.2, Figure 3)**
Randhwa, 1959, p. 332, Figure 82.

Filaments are unbranched, cells are cylindrical, vegetative cells are longer than the broad, chloroplasts 3, spirally arranged, parietal ribbon like bodies with numerous prominent pyrenoids. The nucleus is centrally situated in a protoplasmic strand.

Floating; Collection no. SA (7); Collection site: Palbani rice field; pH 6.8; Date of collection: 03.04.2006.

***Closterium dianae* Ehrenberg (Plate 10.2, Figure 4)**
Tomaszewicz, 1974, p.402, Plate 11; F.4.

Elongated spindle shaped thallus tapering towards both ends from the middle, absence of constriction in the isthmus region, presence of small vacuole at each end.

Submerged floating; Collection no. S7 (i); Collection site: Rajabasa rice field; pH 7.0; Date of collection: 20.03.2006.

***Cosmarium cyclium* Lund. (Plate 10.2, Figure 5)**
West and West, 1905, p.146, Pl.58, Figure 10.

Cells are small, constricted into two symmetrical semi cells joined by an isthmus. The cell wall is two layered, two chloroplasts

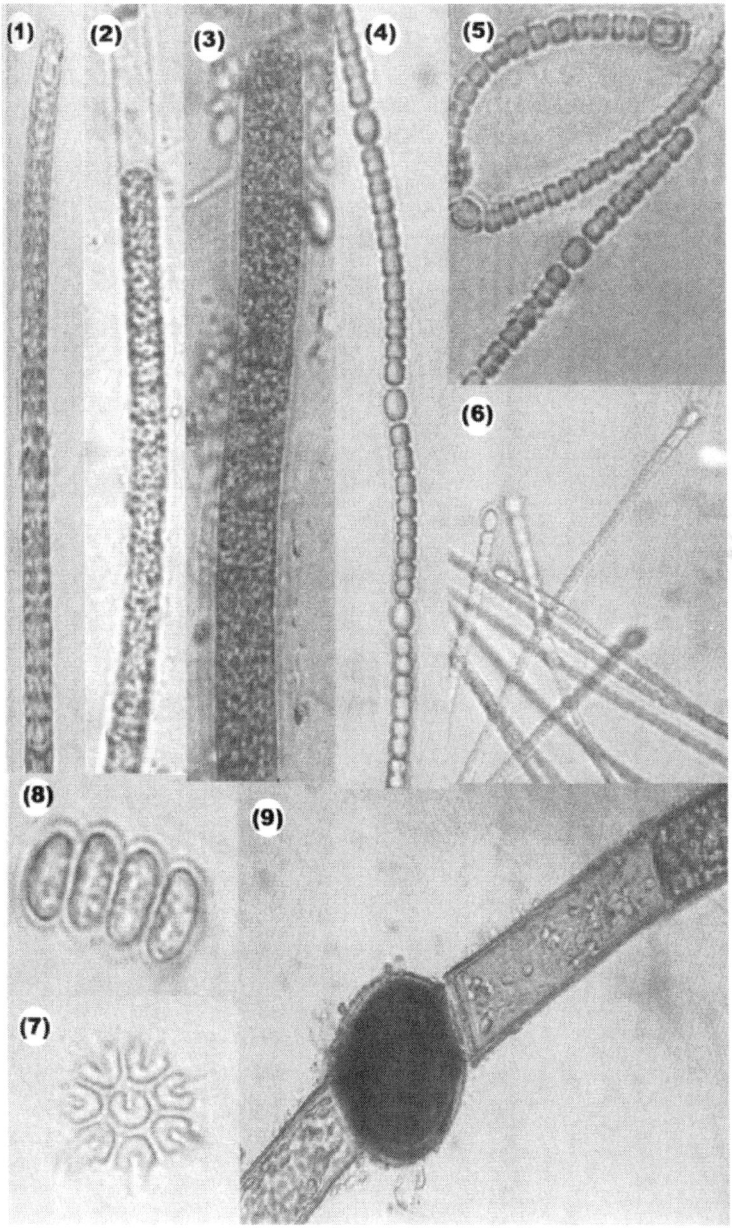

Plate 10.1

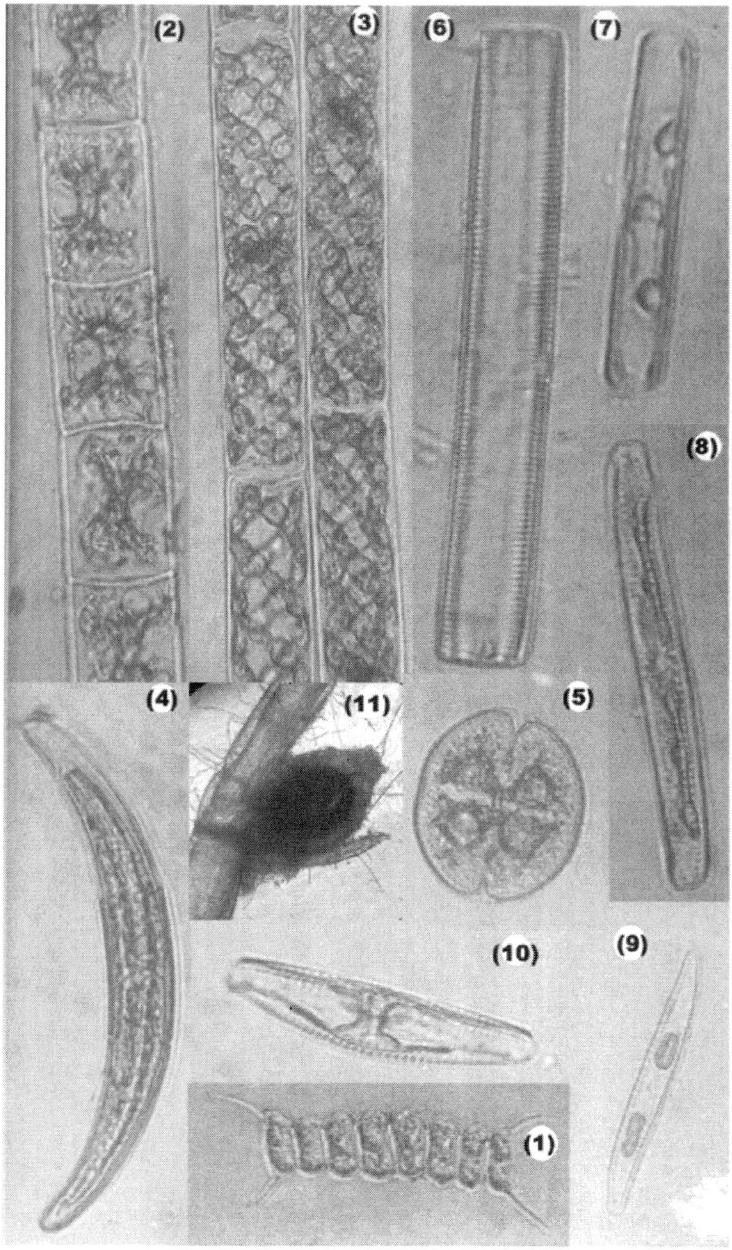

Plate 10.2

in each semi cell, the chloroplasts contain pyrenoids, the cells are uninucleate.

Submerged floating; Collection no. SB (xiii); Collection site: Karatabasa, rice field; pH 6.8; Date of collection: 06.04.2006.

Bacillariophyceae

Fragilaria sp. (Plate 10.2, Figure 6)

The axial area may give the appearance of a false raphe or pseudo raphe.

Submerged floating; collection no. SB (xiii); Collection site: Karatabasa rice field; pH 6.7; Date of collection: 06.04.2006.

Pinnularia sp. (Plate 10.2, Figure 7)

Isobilaterally symmetrical with a longitudinal slit termed as raphe, raphe is interrupted in the middle by a central nodule formed by internal thickening of the valve, each pole of the valve also contains a nodule (polar nodule).

Floating; Collection no. SB (xiii); Collection site: Karatabasa rice field; pH 6.7; Date of collection: 06.04.2006.

Pinnularia sp. (Plate 10.2, Figure 8)

Isobilaterally symmetrical, apical rib is present, valve marking consisting of fine pores bridges from the inside by sieve membranes containing numerous fine pores.

Submerged floating; Collection no. SB (xiii); Collection site: Karatabasa rice field; pH6.7; Date of collection: 06.04.2006.

Nitzschia sp. (Plate 10.2, Figure 9)

Naked pyrenoid, mucilage is tubular in structure. The half spindle lack striations.

Submerged floating; Collection no. S8; Collection site: Manchabandha rice field; pH 6.8; Date of collection: 15.03.2006.

Cymbella sp. Agardh, 1830 (Plate 10.2, Figure 10)

Cell anulate, with symmetrical valves, presence of trans apical striae.

Free floating; Collection no. SC (8); Collection site: Khetrapatana rice field; pH 6.5; Date of collection: 11.04.2006.

Charophyceae

Chara connivens **Salzmann ex. A. Braun. (Plate 10.2, Figure 11) Pal *et al.*, 1962, p.114-115, Figure 273.**

Plant is dioecious, brittle, though not much incrusted; not known to produce bulbils; stem slender, rather rigid; stem cortex evenly triplostichous, cells of the primary and secondary series of equal prominence; spine cells quite rudimentary; stipuloides in a double whorl, rudimentary.

Fix floating submerged; Collection no. SA (7); Collection site: Sagunasbasa rice field; pH 6.8; Date of collection: 02.03.2006.

Discussion

A total of twenty algal taxa belonging to eighteen genera were collected from rice fields of Baripada and adjoining areas of which six belonging to Cyanophyceae, eight to Chlorophyceae, five to Bacillariophyceae and one to Charophyceae. The predominant Cyanophycean members were *Oscillatoria, Phormidium, Lyngbya, Anabaena, Nostoc* and *Cylindrospermum*. Both heterocystous and non-heterocystous forms were found and most of them occurred in floating form. The common Chlorophyceae were *Scenedesmus, Zygnema, Spirogyra, Closterium, Cosmarium, Pediastrum* and *Oedogonium*. *Scenedesmus* is represented by two distinct species. The common Diatoms identified were members of *Fragilaria, Pinnularia, Nitzschia* and *Cymbella*. *Chara connivens* is the single Charophyceae reported during the survey. The range of pH observed was 6.5–7.2.

Acknowledgement

The authors are thankful to Prof. and Head P.G. Dept. of Botany, North Orissa University, Takatpur, Baripada, Orissa for providing facilities and encouragement. Thanks are due to Dr. S.P. Adhikary, Prof. P.G. Dept. of Botany, Utkal University, Vanivihar, Bhubaneswar for providing laboratory facilities to take microphotographs.

References

Adhikary SP (1998) Cyanobacterial germplasm of Orissa state maintained at the department of Botany, Utkal University. Plant Sci. Res. Vol. 20: 57-63.

Anand N (1998) Indian freshwater Micro-algae. Bishen Singh Mahendra Pal singh, Dehradun. pp.94.

Desikachary TV (1959) Cyanophyta. Indian Council of Agricultural Research, New Delhi, India. pp. 686.

Gonzalves EA (1981) Oedogoniales. Indian Council of Agricultural Research, New Delhi, India. pp. 757.

Mohanty RC and Padhi B (1982) Blue-green algae of Bhubaneswar and adjoining region–II. Phykos 21:90-98.

Mohanty RC and Padhi B (1984) Blue-green algae of Bhubaneswer and adjoining regions-III. Phykos 21: 90-95.

Parija P, Pattnaik H and Rao SD (1969) Blue-green algae of Cuttack. Prakruti- Utkal University, Journal-Science 6:183-186.

Randhwa MS (1959) Zygnemaceae. Indian Council of Agricultural Research, New Delhi, India. pp.478.

Sahu JK, Nayak H and Adhikary SP (1996) Blue-green algae of rice fields of Orissa state-I, Distributional pattern in different agro climatic zones. Phykos 35: 93-110.

Sahu JK (2000) Distribution and monthly succession of blue green algae in the rice fields of Puri district of Orissa. J. Eco. Taxon. Bot. 24: 191-195.

Chapter 11

Impact of Tropical Forest Degradation on Soil Microfungal Diversity

Uma Devi

Reader-in-Botany, Department of Botany,
Govt. Women's College, Sambalpur – 768 001, Orissa

ABSTRACT

Soil microfungal community inhabiting a tropical dry deciduous forest of Orissa was studied. For a comparative assessment of mincrofungal diversity, five forest sites *viz.* a relatively undisturbed forest (natural forest), three comparatively degraded forests and a completely denuded forest were selected on the basis of crown density and vegetation characteristics in Badrama Reserve Forest of Bamra Division. Soil analysis of the natural and degraded sites revealed gradual deterioration of soil quality in terms of physico-chemical parameters indicating impact of vegetation cover loss on soil health. Soil microfungal community was analyzed in terms of species richness, species composition, per cent isolation, dominant species association, similarity index, diversity and dominance indices. A total of 48 species (35 genera), constituting six phycomycetes, three Ascomycetes, 37 Deutromycetes and two sterile mycelia were isolated from the natural forest. Species richness as a whole and with respect to different groups decreased from natural to barren site. Increasing AIV trend of some hardy species indicated their higher adaptability to soil adversity. Dominant microfungal species association (*Aspergillus–Penicillium–Trichoderma*) typical to this forest, changed with the changes in the forest floor. Maximum similarity index was observed between natural and least degraded site and least between natural and barren site because of comparable microclimatic and edaphic factors. Microfungal community diversity in the natural forest showed a declining trend in the degraded sites. A significant positive correlation between overlaying vegetational diversity and microfungal diversity indicated the profound role of canopy vegetation on soil microfungal community.

Keywords: *Forest degradation, Microfungal diversity, Soil physico-chemical parameters, Vegetational diversity.*

Introduction

Forest biochore represents a complex, self–sustaining land ecosystem endowed with a treasure of natural resources harboring a rich bio-diversity. The origin, growth and sustenance of human civilization owe much to this epitome of nature since the dawn of its genesis. As a community, forest exhibits exemplary interaction between flora and fauna against the backdrop of environmental parameters. Soil is one such important parameter that supports and sustains the overlaying vegetation. The thick plant cover of the forest helps to maintain a self–regulatory microclimate optimizing the floor temperature, water contents and soil chemistry (Herbett *et al.*, 1964). Under the suitable physico-chemical setup soil sub-system under the forest canopy harbors a varied population of soil organisms above ground and within, maintaining the detritus food-web that regulates the nutrient dynamics. The soil flora and fauna provide the basic life supporting system that underpin the survival of terrestrial ecosystems (Coleman *et al.*, 1983; Lynch and Bragg, 1985; Tesarova, 1988; Lee and Pankhaurst, 1992). Vegetation of the forest in turn, supports a more fertile and stable soil (Brady, 1988). The vegetational diversity ensures heterogeneity of substrate input to the forest floor thus, becoming an important predictor of the nutrient and fertility status, water balance and over all structure and function of the soil (Lewis and Rice 1990). Structural aspects of soil deal with the physico-chemical parameters and functional aspects refer to the various community characteristics of the soil organism harbored by the forest floor (Wallwork 1976; Herdnix *et al.*, 1990). The biomass of soil biota is more than 20t ha^{-1} ranging from bacteria to earthworm and their role is also diverse as mentioned by Kennedy (1996). Among the diverse pedoflora inhabiting the forest floor, microfungi are ecologically important for their contribution to detritus dynamics (Parkinson and Coleman 1991; Lodge and Cantrell 1995). Fungi are numerically the most abundant soil organisms as reported by Miller and Lodge (1997). Globally saprophytic soil fungal diversity corresponds to 1.5 million with a ratio 1:6 of native plant to fungal species (Hawksworth 1991). As hypothesized by Hammond (1992) the over all forest architecture may be a better predictor of the number of fungi present in a given area than, plant species richness.

Soil disturbance in tropical systems is a common occurrence due to both natural and man made factors (Allen *et al.*, 1998). Tropical dry forests belong to the most threatened ecosystems on earth (Heinrich and Hurka 2004) and tropical forest cover in India is vanishing very fast (Myers 1988). Massive degradation of tropical forest has severely impacted the structure and function of soil underneath (O'Neill *et al.*, 1977; Campbell and Souster 1982) paving way to desertification through the genesis of degraded barren lands (Jurgensen *et al.*, 1996). For the nutrient impoverished tropical soils, the role of soil microbes is essential for the maintenance of viable plant communities (Janos 1980). Ecosystem functioning is largely governed by soil microbial dynamics (Kennedy and Smith 1995). Decrease/alteration in soil biotic community under managed ecosystem as compared to native forest in tropics has been well documented (Lavelle and Pashanasi 1989; Zou *et al.*, 1997). Soil microbial diversity is conformably adopting to the environment and thus, can function as sensitive indicator of changes in soil quality (Kennedy and Papendick 1992) and this can be used to scale the extent of perturbation brought to soil sub-system pertaining to vegetation cover loss. Works of Boyle (1975); Entry *et al.* (1986); Hendrickson *et al.* (1987); Miller and Lodge (1997) report the adverse impact of forest degradation/disturbance on soil microbial diversity and population. It is therefore urgent and important to study and document soil quality with respect to its physical, chemical and biological components to assist the maintenance of tropical land ecosystems. The present study reports an assessment of soil microfungal diversity in response to forest degradation with a view to achieving some means and measures towards conservation of forest.

Study Area

The work was carried out in the Badrama Forest Range under Bamra Forest Division of undivided Sambalpur district of Orissa. The area is located between 21°05′ to 21°12′ North latitude, 83°13′ to 84°08′ East longitude and 300m altitude. Badrama forest constitutes one of the physiographic sections of the Eastern Ghats. The whole tract is hilly interspersed by valleys. The reserve forest starts at a distance of 21.5 km from Sambalpur headquarters and flourishes along N.H.6. Climate of this area is monsoonal with three seasons *viz.* monsoon (June–September) winter (October–January) and summer (February–May). Average annual rain fall received by the area is 1533mm (average of twelve years 1986-1997 recorded in the Hirakud Research Station, Sambalpur), 90 per cent rainfall being due to monsoon. The mean monthly air temperature ranged beween15°C (January) to 43°C (May). Relative humidity averaged between 46-85 per cent. Soil of this area is brownish derived from Khandolites and Charnokites. Soil is sandy loam type and slightly acidic in reaction.

Badrama Forest is an important forest of western Orissa being the chief source of valuable timber species like sal, bijasal and rosewood and wide varieties of non-timber forest produce (NTFP). Thus, this forest helps augment of the state revenue. Supporting a rich diversity of wild fauna recently it has been declared Badrama Wild Life Range under Badrama Wildlife Division. It houses the famous Ushakothi National Wildlife Sanctuary. The forest of this range has been facing severe degradation for last five decades. Unabated and unscientific harvest of timber and NTFP, clear felling, overgrazing, clearing of forest for village farming and human settlement are causes of forest fragmentation and degradation. Further, irregular monsoon and repeated forested fire cause severe damage to the seeds and seedlings and aggravate the soil of the forest. Keeping these facts in view the present study was done to asses the impact of forest degradation and loss of vegetation cover on soil and soil mycoflora. Outcome of such studies have relevance for the formulation of appropriate management strategy as a step towards protection and conservation of forest.

Study Sites

The survey was undertaken in and around Badrama forest within 20 km radius. On the basis of vegetation characteristics and estimated canopy density (zero to one scale) five sites were selected as follows (Presented in Plates 11.1–11.5).

Site–I

A relatively undisturbed forest. This site displayed a luxuriant vegetation with 0.8 canopy density. Subsequently this site will be referred to as natural forest (NF).

Site–II

Degraded forest (DF_1): Canopy density was 0.6.

Site–III

Degraded forest 2 (DF_2): Tree stands were widely located interspersed by coppices and shrubs, canopy density was 0.3.

Site–IV

Degraded forest 3 (DF_3): A degraded savannised scrub with few stunted trees and bushes. Floor was eroded with patchy undergrowths.

Site–V

Barren Site (B): This site of the forest was completely devoid and canopy vegetation, floor was covered with seasonal undergrowths.

Vegetation Characteristics of the Sites

The climatic vegetation of Badrama forest corresponds to Northern tropical dry deciduous type displayed in three tiers (Champion and Seth 1968). Top storey consists of tree species (height 15 to 25 mm), middle–storey with tree species (height 10 to 15m). Third tier is formed of shrubs and herbs. Liana and herbaceous climbers form a prominent life form of the forest. Besides, orchids, loranths, ferns and mosses constitute the minor flora of the forest. A phytosociological survey undertaken adopting random sampling procedure in five hectare plots at each study site, reveled the presence of 120 angiospermic plant species belonging to 46 families in the natural forest. The life forms constituted 55 tree species (girth class >

Plate 11.1: Natural Forest (NF)

Plate 11.2: Degraded Forest 1 (DF_1)

Plate 11.3: Degraded Forest 2 (DF_2)

Plate 11.4: Degraded Forest 3 (DF_3)

Plate 11.5: Barren Site (B)

15cm gbh) belonging to 25 families, 18 shrub species (in 13 families) 20 species of liana (13 families) and 27 herb species (15 families). *Shorea robusta–Terminalia tomentosa, Ptetrocarpus marsupium* formed the dominant species association which agrees with earlier record for forest types of western Orissa (Puri *et al.*, 1989) and this forest represents a dry peninsular sal forest of India. Vegetation analysis of the five sites revealed that species richness and stand density of trees, shrub and liana so also tree basal area declined from natural to barren site. The species association typical to this forest region, exhibited a shift in the degraded site with prominence of scrub-land species such as *Soymida febrifuga, Cleistanthus collinus, Semecarpus anacardium*, *Buchanania lanzan* and *Butea monosperma* (Devi and Behera 2003). Prominence of these species is reported to be an indicative symptom of dry deciduous forest degradation (Panigrahi 1970). Increase of invading non-forest herb species, their seasonal increase in density due to open canopy further scale the degradative state of the forest.

Methods

Soil Sampling

Ten soil samples from each study site were collected from randomly dug 10 × 10 × 15 cm pits and the samples were composited in the laboratory to make a representative sample for each site. The samples were kept in sterile poly packets. From each composite sample one part was used to analyze soil physico-chemical parameters (Brady 1988). The other halves were kept in field moist condition for culture of microfungi. Soil temperature and soil reaction (pH) were recorded in the field. Field view of the vertical sections of the soil were taken for soil profile study (Brady 1988).

Microfungal Culture and Isolation

Soil samples were collected at monthly intervals for a period of 18 months (from January 1996 to June 1997). Five replicate culture petriplates for each composite samples were run for the culture, isolation and enumeration of soil microfungi, adopting dilution plate count method (Waksman 1922). This method as reported by Christensen (1969) draws sample population from greater volume of soil than do most other methods and therefore species restricted to small pieces of organic or mineral particles are not so apt to be missed. For mixed and pure culture Martin Rose Bengal Streptomycin Agar media (Martin 1950) was used. The whole procedure was run on the hot air laminar flow (MAC). Fungal colonies were examined and counted at an interval of three days, till 21 days allowing slow growing species to grow. One set of culture plates without inoculum served as control. For preservation, agar slants were prepared from pure culture using Czapek Dox Agar media (Thom and Raper 1945). Microfungal species isolated during the study were identified using books/monographs by: (i) Barnett HL and Hunter B *Illustrated General of Fungi Imperfecti*, Burgess Publishing Co. 1972; (ii) Booth C. *The genus Fusarium*, Common Wealth Agric Bureaue, England 1971 and (iii) Gilman GC. *A manual of soil Fungi*. Oxford and IBH Publishing 1967.

Enumeration of Microfungal Population

Number of colony forming units (CFU) from five replicate petriplates (per unit composite sample) were counted for each microfungal species taking dilution factor into consideration.

Per cent isolation was calculated using the formula.

$$\text{Per cent isolation} = \frac{\text{No. of CFU of individual species in 5 replicate petriplates}}{\text{Total no. of CFU of all species in 5 replicate petriplates}} \times 100$$

(Cavender and Raper 1965)

Per cent isolation of individual species was taken as importance value index (IVI). Shannon's diversity index (H) and Simpson's dominance index (Cd) were calculated following Magurran (1988).

Results and Discussion

Soil Characteristics

The physico-chemical characteristics of soil from the natural and degraded sites (Table 11.1) revealed gradual decrease of clay fraction, soil porosity water holding capacity, soil moisture content and increase of bulk density, soil temperature variability from natural to degraded sites. Organic carbon, total nitrogen level also followed the trend NF>DF_1 > DF_2> DF_3 > B. Soil reaction in terms of pH declined from nearly a neutral range (6.7) in the natural forest to 5.8 indicating gradual acidification of soil which agrees with Balagopan (1995) and Rhoades and Binkley (1996) that savannisation and substitution of natural forest results in acidification of soil. Soil profile analysis reflected the reduction of surface litter accumulation and gradual thinning of organic horizons in the degraded to barren sites (Devi 2001) giving the visual symptom of soil deterioration. Soil colour is considered an indirect index of soil fertility (Mitra *et al.*, 1980). A sharp colour gradation (Plate 11.6) from dark-black soil of natural forest to grey soil of the barren site indicated poor fertility status of the degraded sites.

Table 11.1: Physico-chemical Characteristics of Soil in Natural and Degraded Forest Sites

Sl.No.	*Parameters*	*NF*	*DF_1*	*DF_2*	*DF_3*	*B*
(i)	Soil separates (%)					
	Sand	76.0±0.79	78±0.48	79.75±0.62	81.0±0.35	80±0.35
	Silt	7.0±0.66	7.5±0.07	7.5±0.95	8.0±0.35	9.4±0.77
	Clay	17.0±0.69	14.5±0.78	12.75±0.36	11.0±0.65	10.6±0.19
(ii)	Bulkdensity (cm^3)	1.11±0.07	1.21±0.09	1.39±0.01	1.50±0.01	1.62±0.03
(iii)	Porosity (%)	58.01±2.85	54.34±2.52	47.75±0.16	43.39±0.70	38.47±1.07
(iv)	Water-holding capacity (%)	48.38±0.7	44.0±0.62	39.77±0.31	33.92±0.67	31.47±1.07
(v)	Soil pH	6.7±0.24	6.53±0.12	6.41±0.14	6.12±0.27	5.86±0.35
(vi)	Soil temperature variability range	26±0–32.5±0.16	26.8±0.26–35.5±0.39	26.8±0–38.8±0.37	28.7±0.4–41.5±0.97	28.3±0.41–42.5±0.44
(vii)	Organic carbon (%)	2.70±0.02	2.08±0.3	1.82±0.4	1.11±0.4	0.94±0.31
(viii)	Total nitrogen (%)	0.246±0.02	0.202±0.01	0.141±0.015	0.183±0.02	0.134±0.014

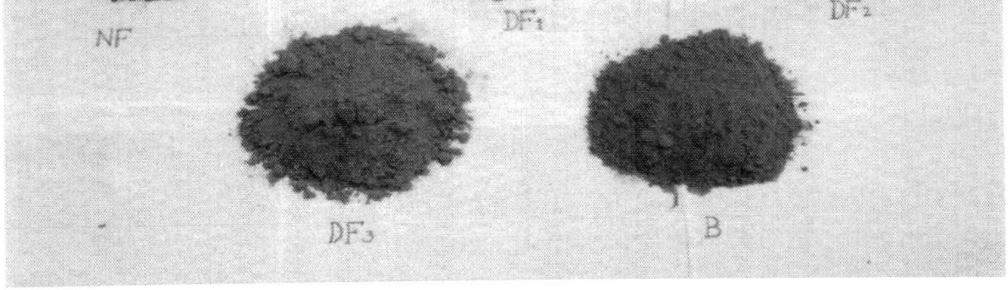

Plate 11.6: Colour of Soil Collected from Natural and Degraded Forest Site

Microfungal Community Characteristics

Species Richness

Forty eight species of microfungi belonging to 35 genera were isolated from natural forest soil representing, six Phycomycetes (12.5 per cent), three Ascomycetes (6.25 per cent), 37 Deuteromycetes or Imperfecti Fungi (77.08 per cent) and two sterile mycelia (4.17 per cent). Abundance of deutermycetous species and relatively low isolaton of ascomycetous species from tropical dry forest soil of Orissa (Behera 1980) and different forest types of India (Dwivedi 1966; Shukla *et al.*, 1978; Upadhyay and Rai 1979) have already been reported. The genus *Aspergillus* was isolated with 11 species. The dominance of Aspergilli in dry deciduous acidic forest soil is in confirmation with the views of Jensen (1951); Farrow (1954); Kelleman and Mcbeth (1972) and Chakravarti *et al.* (1991).

The other common fungi isolated were *Penicillum* with (three species) and *Fusarium* with (two species). Rest 45 genera isolated were monospecific. The two sterile mycelia isolated were white and black mycelia. Comparison revealed that number of species and genera isolated from natural to degraded sites followed a declining trend (Figure 11.1).

Microfungal Species Composition and Per cent Isolation

Table 11.2 complies data of average per cent isolation value (AIV = average of 18 sampling months) of microfungal species and Figure 11.2 presents AIV of different microfungal groups isolated from natural to degraded sites. The phycomycetous species reported to be moisturephilic (Dwivedi 1966; Mishra and Kanaujia 1972) responding positive to the soil moisture regime. As observed, species composition of phycomycetes group as a whole declined from natural to barren site with complete disapperance of species of *Absidia, Actinomucor* and *Syncephalastrum* in the degraded sites. (Table 11.2). *Cunninghamella* though isolated from all the sites its AIV declined from natural to barren site. On the contrary, species of *Rhizopus* and *Mucor* showed increasing AIV which might have added to higher AIV of phycomycetous group in the barren site. This shows relatively greater adaptability and tolerance of *Mucor* and *Rhizopus* to soil adversity (Alexopoulos 1978). In Ascomycetes, species composition, AIV of individual species as well as and AIV of the group as a whole declined from natural to barren site.

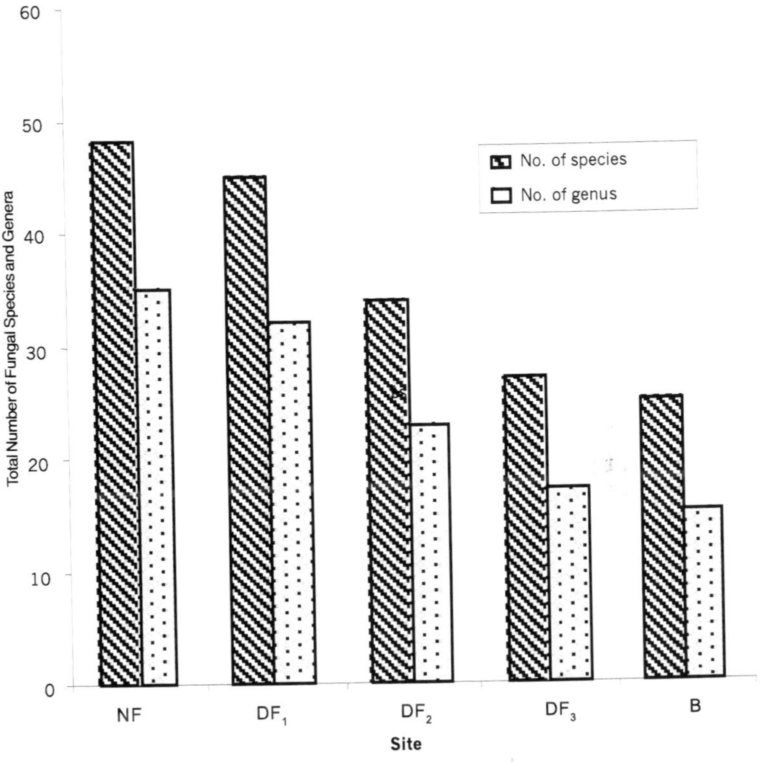

Figure 11.1: Total Number of Microfungal Species and Genera Isolated from Natural and Degraded Forest Sites

It could be concluded that tropical forest degradation exerts much impact on this group of microfungi. Out of 37 species of Fungi Imprefecti isolated from Badrama forest soil 23 species (62 per cent) were common to natural as well as degraded and barren sites. Genera such as *Aspergillus, Penicillium, Alternaria, Curvularia, Helminthosporium* showed increased AIV from natural to barren site adding to an increasing AIV trend of Deuteromycetes from natural to degraded sites. Their high dominance may be ascribed to wide tolerance of climatic and edaphic stresses (Warcup 1950, Singh and Charya 1975).

Figure 11.2: AIV of Microfungal Groups Isolated from Natural an Degraded Forest Sites

Phycomycetes

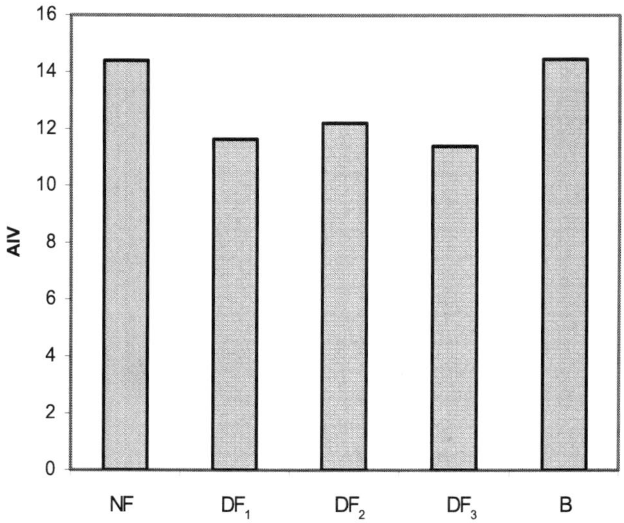

Ascomycetes

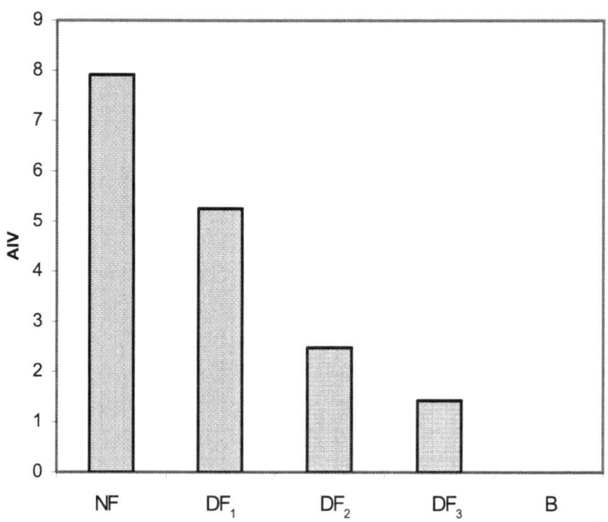

Contd...

Figure 11.2–Contd...

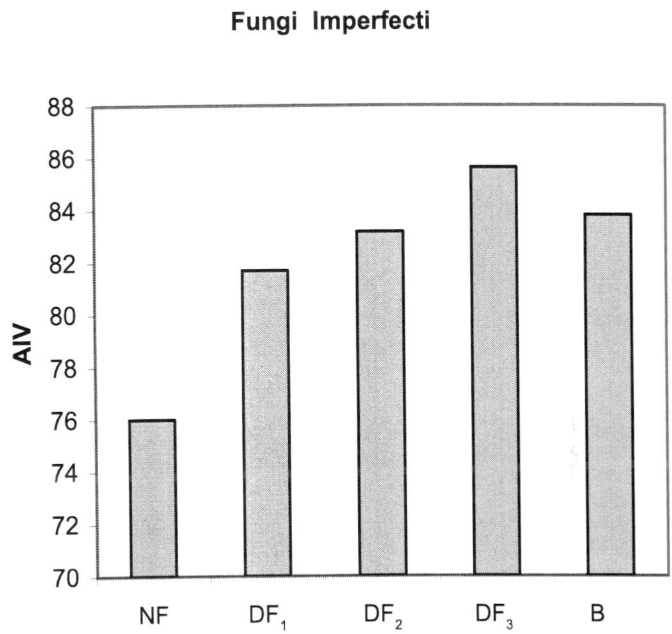

Contributing highest AIV, *Aspergillus* was the dominant genus followed by *Penicillum* and *Trichoderma*. AIV of these three genera has been presented in Figure 11.3. Though number of species of *Aspergillus* and *Penicillium* declined from natural to barren site the AIV of these two genera increased which indicates that forest degradation has its adverse impact on their qualitative composition. Species of *Aspergillus* such as *A. riger, A. fumigatus, A.clavatus, A. flavus* and two species of *Penicillium* were found to be hardy. Tiwari and Rai (1977); Upadhyaya and Rai (1979); Behara *et al.* (2000) have reported wide-range edaphological adaptability of these species of *Aspergillius* an *Penicillium*. Greater occurrence of *Trichoderma* in NF and DF_1 with a subsequent decline in the rest degraded sites was marked. This finding corroborates the fact that the two factors, such as cool environment and humus rich soil regulate the distribution of *Trichoderma* (Danielson and Davey 1973; Sharma *et al.*, 1995). With forest degradation these two factors are adversely effected for which *Trichoderma*, reported as a common forest fungus (Bisby *et al.*, 1935; Chesters 1949) is expected to show low isolation.

Table 11.2: Complied Data of Microfungal Species Isolated from Different Sites with their Average Per cent Isolation Values (AIV)

Microfungal Species	NF	DF_1	DF_2	DF_3	B
PHYCOMYCETES					
Absidia sp.	1.16	0.68	–	–	–
Actinomucor repens	1.19	1.00	–	–	–
Cunninghamela ecbninulata	3.62	2.26	1.69	1.23	0.54
Mucor hiemalis	1.78	1.70	2.88	3.88	4.35
Rhizopus nigricans	3.87	4.13	6.87	6.30	9.57
Syncephalastrum racemosum	2.80	1.87	0.77	–	–
ASCOMYCETES					
Chaetomium globosum	2.42	1.47	–	–	–
Colletotrichum sp.	4.24	3.79	2.49	1.43	–
Thielavia terricola	1.26	–	–	–	–
DEUTEROMYCETES					
Moniliaceae					
Acremonium vitis	0.46	0.31	–	–	–
Aspergillus candidus	2.16	1.67	2.63	2.83	1.54
Aspergillus clavatus	2.97	4.98	5.23	3.75	6.92
Aspergillus flavus	3.14	4.36	5.29	4.01	4.32
Aspergillus fumigatus	5.11	3.81	6.28	11.91	10.01
Aspergillus luchuensis	1.21	0.80	–	–	–
Aspergillus nidulans	1.95	1.23	–	–	–
Aspergillus niger	6.52	6.53	8.91	9.02	10.68
Aspergillus repens	1.75	0.98	1.25	1.48	1.58
Aspergillus sydowi	1.08	0.90	2.25	5.18	5.41
Aspergillus terreus	2.59	3.54	3.21	1.49	1.76
Aspergillus ustus	1.02	1.74	0.49	–	–
Geotrichum candidum	1.91	2.09	0.80	1.38	0.37
Gliocladium sp.	1.18	1.96	0.68	–	–
Paecilmyces sp.	2.51	1.84	1.56	1.28	0.99
Penicillium steckii	4.23	3.54	5.31	5.85	6.60
Pencillium sp1	1.98	2.74	5.31	4.61	3.58
Penicillum sp 2	1.26	1.84	2.69	4.77	4.78

Contd...

Table 11.2–Contd...

Microfungal Species	NF	DF_1	DF_2	DF_3	B
Spicaria sp.	1.02	1.45	–	–	–
Tricboderma viride	5.40	4.92	4.82	2.16	1.68
Verticillium sp.	1.90	2.64	3.72	1.04	3.73
Dematiaceae					
Aureobasidium sp.	0.47	–	–	–	–
Alternaria alternata	2.63	3.33	3.27	3.25	3.74
Botryotrichum sp.	1.60	1.03	0.46	–	–
Cladosporium herbarum	2.85	3.13	1.05	–	–
Curvularia lunata	2.95	4.24	4.45	11.42	8.97
Gliomastix sp.	1.77	1.40	0.66	–	–
Helminthosporium sp.	1.88	2.66	1.96	2.48	3.51
Humicola grisea	2.60	2.47	3.02	2.19	1.09
Memnoniella echinata	0.59	–	–	–	–
Nigrospora sphaerica	1.98	1.45	–	–	–
Papularia sphaerosperma	0.28	0.45	–	–	–
Stemphylium veruculosum	0.69	0.67	–	–	–
Torula sp.	1.69	0.84	–	–	–
Tuberculariaceae					
Fusarium culmorum	1.36	3.59	4.12	2.90	2.56
Fusarium sp.	0.74	1.86	2.79	1.72	–
Hymenula affinis	0.56	0.80	0.95	–	–
Strile mycelia					
White mycelia	1.34	0.67	1.84	0.68	1.21
Black mycelia	0.36	0.61	0.42	0.94	0.46

Dominant microfungal association

On the basis of AIV, the sequence of dominant genera in the natural and degraded forest sites observed were:

NF: *Aspergillus–Penicillium–Trichoderma–Colletotrichum–Rhizopus*

DF_1: *Aspergillus–Penicillium–Trichoderma–Curvularia–Rhizopus*

DF_2: *Aspergillus–Penicillium–Rhizopus -Trichoderma–Curvularia*

DF_3: *Aspergillus–Penicillium–Curvularia–Rhizopus–Mucor*

B: *Aspergillus–Penicillium–Rhizopus–Curvularia–Mucor*

Figure 11.3: AIV of Dominant Microfungal Genera Isolated from Natural and Degraded Forest Sites

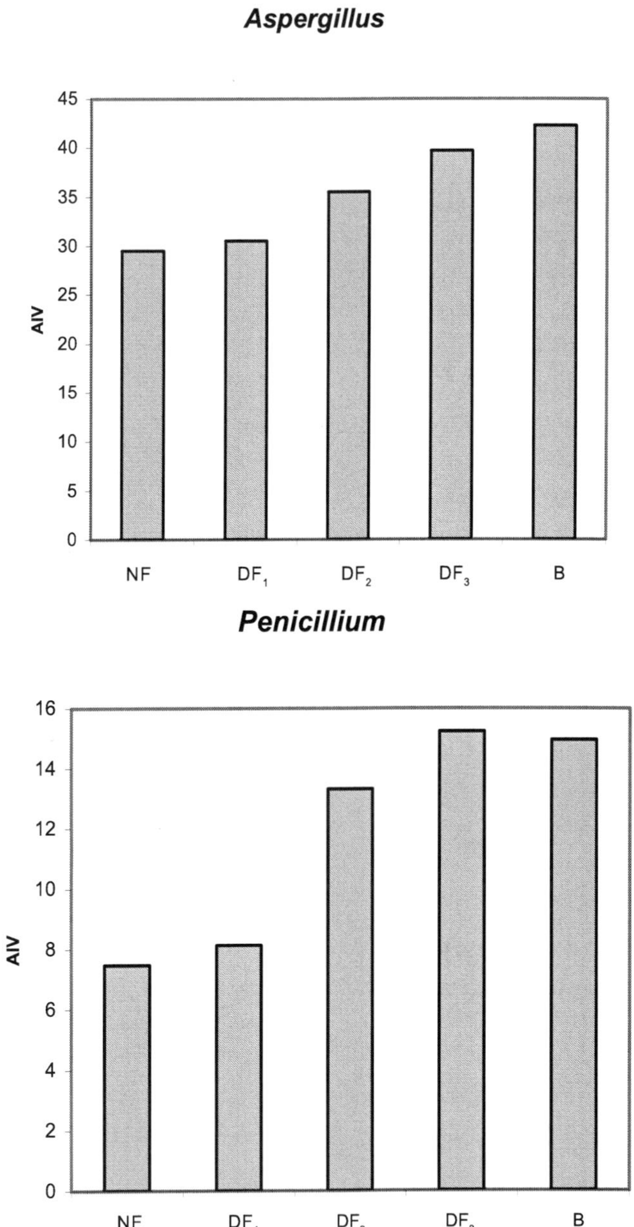

Contd...

Figure 11.3–Contd...

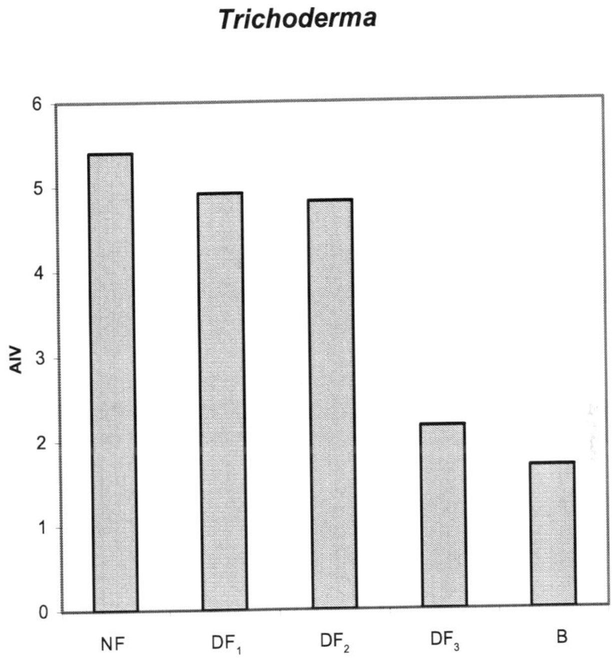

The above shift in species association indicated change in fungal community structure as a result of change in the forest cover. This may have implications on the soil function on the basis of the fact that the soil with altered microfungal community may have alteration in the community function.

Similarity Index (s) of Microfungal Species

Similarity index of microfungal species with respect to different sites were calculated following Sorensen (1941)

$$S = 2j/(a+b)$$

Where,

a: Species exclusive to site A

b: Species exclusive to site B

j: Species common to site A and B sites.

	NF	*DF_1*	*DF_2*	*DF_3*	*B*
NF	–	30	5.38	2.80	2.17
DF_1	–	–	7.0	3.29	2.50
DF_2	–	–	–	8.0	5.0
DF_3	–	–	–	–	16.66
B	–	–	–	–	–

Analysis of similarity index with respect to fungal species composition showed maximum similarly between natural and least degraded forest (DF_1), least similarity between barren site and NF and relatively high similarity between most degraded (DF_3) and barren site. Comparable micro-climatic and edaphic factors can be considered to be reasons for such similarity in the microfungal community structures between different sites. On the contrary, variation of such factors between different sites may explain the dissimilarity of the community structure.

Shannon's Diversity Index (H) and Simpson's Dominance Index (Cd) of Microfungal Species

Micro-fungal community diversity and dominance indices revealed a decrease in diversity value and increase in dominance value from natural to degraded site. Thus the relationship between the two indices was of inverse order. This supports the ecological concept that for a diversified community, dominance is only restricted to few species (Magurran 1988). This study further revealed that with the progress of forest degradation microfungal diversity showed a declining trend.

Correlation Between Vegetational Diversity and Microfungal Diversity

A significant positive correlation of microfungal diversity was observed with the total overlaying plant diversity, $r = 0.745$ ($P < 0.01$) and diversity of tree species $r = 0.968$ ($P<0.001$). This indicates that natural forest cover having varieties of plant species, promotes microfungal diversity in soil (Jha *et al.*, 1992; Henrot and Robertson 1994). Christensen (1969) working on the coniferous forest of Wisconsin reported that species composition within the micro-fungal community is closely related to the species composition of the over-laying vegetation. Behera (1980) in a comparative

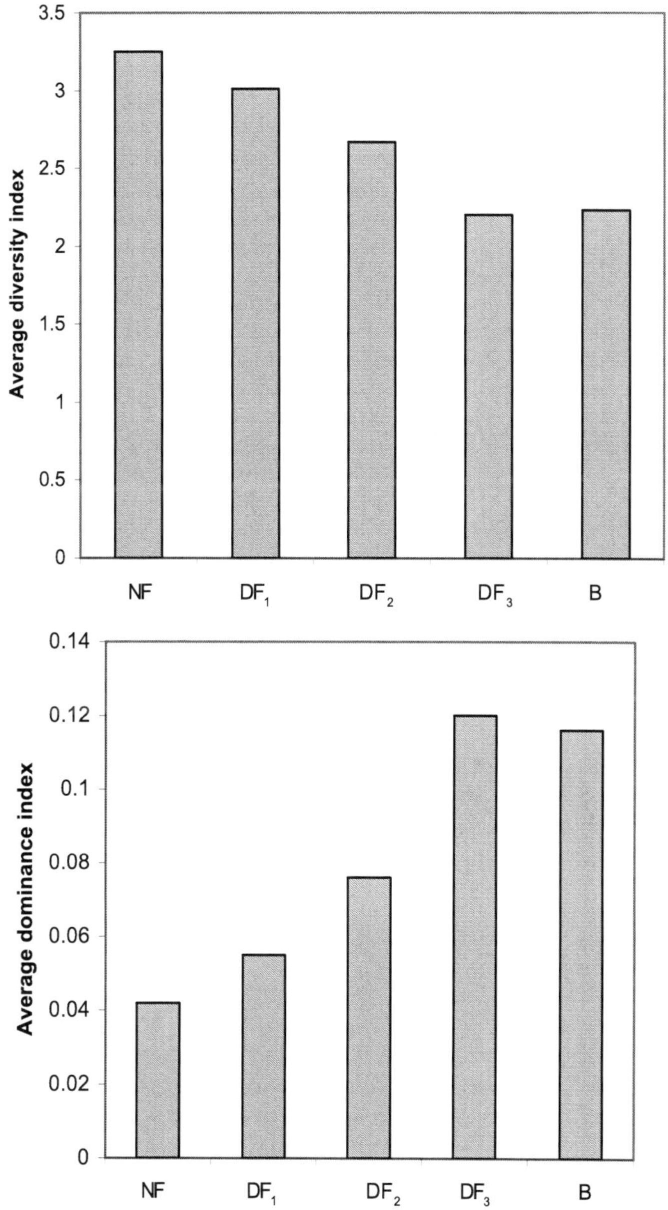

Figure 11.4: Average Diversity Index and Dominance Index of Microfungal Species Isolated from Natural and Degraded Forest Sites

investigation on soil micro fungi of tropical soils of crop field, grass land and forest, emphasized that with the overlaying floral diversity of natural forest, there is heterogeinity with respect to root exudates and plant substrate and this chemical diversity of soil imparts profound role for higher micro-fungal diversity and greater population in forest soil. There exists a close affinity between vegetation cover, plant litter quality and soil-biological attributes in terms of soil microfungal composition and activity in forest ecosystem (Alexander 1978 Wardle and Lavelle 1997) and this a whole help the forest to grow as natural ecological system.

Soil is the resource critical to the maintenance and conservation of eco-system. For the betterment of eco-system like forest, it is therefore, imperative to increase our understanding of soil and soil biological components specifically the decomposer community.

Acknowledgements

The author wishes to thank to Dr. N. Behera, Prof. School of Life Sciences, Sambalpur University for his valuable suggestions and guidance during the work.

References

Alexander M (1978) Introduction to soil microbiology. Widely Eastern Ltd. 467 p.

Alexopoulos CJ (1978) Introductory Mycology. Wiley Eastern Ltd. 467 p.

Allen FB Rincon E, Allen MF, Perez-Jemenez A and Huante P (1998) Disturbance and seasonal dyanamics of mycorrhizae in a tropical deciduous forest in Mexico. Biotropica 30: 261-274.

Balagopalan M (1995) Soil chemical characteristics in a natural forest and adjacent exotic plantation in Kerala. Journal of Tropical Forest Science 8 (2) 161-166.

Behera N (1980) Ecology of soil fungi in some tropical soils of Orissa, India. Ph.D. Thesis, Sambalpur University. 139 p.

Behera N., Devi U, Sahani U and Pattnaik S (2000) Soil microfungal community in *Eucalyptus* plantation and natural forest. Journal of Science and Technology XII (A): 23-32.

Bisby GR, Timonin M and James N (1935) Fungi isolated from soil profile in Manitoba Can. J. Res. Bot. Sci.13: 47-65.

Boyle JR (1975) Whole tree harvesting: Impact on site quality In: Proceedings on TAPPI Annual Meeting, Newyork, 24-26 Feb. Technical Association of pulp an paper industries Atlanta pp.29-32.

Brady N C (1988) The nature and properties of soils. Eurasia Publishing House 150 p.

Campbell CA and Souster W (1982) Loss of organic matter and potentially minerizable N from Saskatchewan soils due to cropping. Can J. Soil Sci. 62: 651-656.

Cavender J C and Raper K B (1965) The Acrasieae in nature. Amer, J.Bot. 52: 294-296.

Chakravarti L, Panwar S and Shukla RV (1991) Effect of closure on soil properties and its fungal population in sal forest: J. Trop. For. 7(1): 51-61.

Champion H G and Seth S K (1968) A Revised Survey of Forests of India. Manager of Publications, Govt. of India, Delhi, 404 p.

Chesters CGC (1949) Concerning fungi inhabiting soil. Trans. Brit. Mycol. Soc. 32: 197-216.

Christensen M (1969) Soil fungi of dry mesic conifer hardwood forests in Northern Wisconsin. Ecology 50:9-27.

Coleman DC, Edward AL and Cole VC (1983) Biological strategies of nutrient cycling in soil systems. Adv. Eco. Res. 13: 1-55.

Danielson RM and Davey CB (1973) The abundance of *Trichoderma* propagules and the distribution of this species in forest soils. Soil Biol. Biochem. 5: 485-494

Devi U (2001) Impact of forest degradation on soil microfungal community. Ph.D. Thesis Sambalpur University 219 p.

Devi U and Behera N (2003) Assessment of plant diversity in response to forest degradation in a tropical dry deciduous of Eastern Ghats in Orissa. Journal of Tropical Forest Science 15 (1): 147-163.

Dwivedi RS (1966) Ecology of some soil fungi of some grassland of Varanasi II. distribution of soil mycoflora. Tropical Ecology 7: 84-89.

Entry JA, Stark NM and Loewenstein H (1986) Effect of timber harvesting on microbial biomass fluxes in a northern rocky mountain forest soil. Can. J. For. Res. 16: 1076-1081.

Farrow WM (1954) Tropical soil fungi Mycologia 46: 632-646.

Hammord PM (1992) Species Inventory In: Global biodiversity status of Earth's living resources. Ed. Groombridge G., Chapman and Hall, London. 17-39pp.

Hawksworth DL (1991) The fungal dimension of biodiversity: magnitude, significance and conservation, Mycol. Res. 95: 641-655.

Heinrich A an Hurka H (2004) Species richness and composition during sylvigenesis in a tropical dry forest in northwestern Coasta Rica. Tropical Ecology 45(1): 43-57.

Hendrickson OQ, Burgess DM and Chatarpaul I (1987) Biomass and nutrients in great Lakes St. Lawrence forest species: Implication for whole tree conservational harvest. Can. J. For. Res. 17:210-218.

Hendrix PF, Crossley DA Jr. Blair JM and Coleman DC (1990) Soil biota as components of sustainable agro-ecosystems. In: Sustainable agricultural systems Ed. Edward CA *et al.* Soil and Water Conservation Society, Ankey, Iowa.

Henrot J and Robertson GP (1994) Vegetation removal in two soils of the humid sub-tropics: Effect on microbial biomass. Soil Biol. Biochem. 26: 111-116.

Herbett CS; Hobba RL an Marvin RJ (1964) Hydrology of forest lands and range lands In: Hard book of applied hydrology (ed) Van Te Chow, McGraw Hill 22-52 pp.

Janos DP (1980) Vesicular arbiscular mycorrhizae affect low and tropical rain forest plant growth. Ecology 61: 151-162.

Jensen HL (1951) The fungus flora of Soil. soil Sci. 123-138.

Jha DK, Sharma GD and Mishra RR (1992) Ecology of soil microflora and mycorhizal symbionts at degraded forests at two altitudes. Biol. Fertil. Soils. 12 272-278.

Jurgensen MF, Harrey AE, Graham RT, Page Dum roese DS, Tonn JR, Larsen MJ and Jain TB (1996) Impact of timber harvesting on soil organic matter, Nitrogen productivity and health Inland, Northwest forest. For. Sci.43 (2): 234-240.

Kellerman AE and McBeth IC (1972) The fermentation of cellulose. Zbl. Bukt. II Abt.34: 485-494.

Kennedy AC (1996) Soil microbial diversity in agricultural systems. In: exploring role of diversity in sustainable agriculture (Olson RK, Francis CA and Kaffka (eds) ASA Madison, WI 35-54pp.

Kennedy AC and Papendick RI (1992) Microbial characteristics of soil quality. J. Soil water conservation. 245-298.

Kennedy AC and Smith KL (1995) Soil microbial diversity and sustainability of agricultural soil. Plant and soil 170: 75-86.

Lavelle P and Pashanasi B (1989) Soil microfauna and management in Peruvian and Amazorian (Yurimaguas, Loreto) Pedologia 33: 233-291.

Lee KE and Pankhaurst CE (1992) Soil organisms and sustainable productivity. Aust. J.Soil. Res.30:855-892.

Lewis J and Rice RM (1990) Estimating erosion risk on forest lands using improved methods of discriminant analysis. Water Resources Research 26: 1722–1733.

Lodge DJ and Cantrell S (1995) Fungal community in wet tropical forests: Variation in time and space. Can J. Bot. 73 (Suppl.) S1391–S1398.

Lynch JM and Bragg E (1988) Microrganisms and soil aggregate stability. Adv. Soil. Sci. 2: 113-171.

Magurran AE (1988) Ecological diversity and its measurement. Princeton Univ. Press. New Jersey, pp 145-146.

Martin JP (1950) Use of acid rose bengal and streptomycin in the plate method for estimating soil fungi. Soil Sci. 69: 215-232.

Miller RM and Lodge JD (1997) Hierarchy of root and mycorrhizal fungal interactions with soil aggregation. Soil Biol. Biochem. 22: 579-584.

Mishra PR and Kanaujia RS (1972) Soil fungi in relation to varying cover vegetation, climatic conditions, soil depth and physico-chemical characters of soil. Proc. Nat. Acad. Sci. India. XLII (B): 131-140.

Mitra GN, Sahu D and Sahu GC (1980). A laboratory manual of soil and water testing. OUAT, Bhubaneswar, 148 p.

Myers N (1988) Threatened Biotas, "Hot Spots" in tropical forests. The Environmentalist. 8: 187-208.

O' Neill RV, Ausmus BS, Jackson DR, Van Hork RI, Van Voris P, Washbaourne C and Watson AP (1977) Monitoring terrestrial ecosystem by analysis on nutrient export. Water, Air and Soil Poll. 271-277.

Panigrahi SS (1970) Revised working plan for the reserve forests of Bamra. Govt. of Orissa. Publ. 474 p.

Parkinson D and Coleman DC (1991) Methods for assessing soil microbial population, activity and biomass. Agril. Ecosyst. Environ. 34: 3-33.

Puri SK, Gupta RK, Meher Homji VM and Puri S (1989) Forest. Ecology Volume 2. Oxterd and IBH Publishing 582 p.

Rhoades C and Binkely D (1996) Factors influencing declining in soil pH in Hawaian *Eucalyptus* and *Albizia* plantation. For. Ecol. And Manag. 80: 225-238.

Sharma GD, Mishra RR and Kshatriya S (1995) Fungi and litter decomposition in the tropics In: Soil organisms and decomposition in the tropics Ed. Reddy MV. pp.39-57.

Shukla AN, Tandon RN and Gupta RC (1978) Phyllosphere mycoflora colonizing the leaf litter of sal (*Shorea robusta*, Gaertn.) in relation to some environmental factor. Tropical Ecology 19 (1): 1-6.

Singh PN and Charya MV (1975) Soil fungi of a sugarcane field at Meerut. I. Distribution of soil mycoflora. Geobios 2: 40-43.

Sorensen T (1941) Temperature relation and phenology of the North East Greenland flowering plants. Meddel Greenland pp.125–305.

Tesarova M (1988) Micro-organisms and Carbon cycle in terrestrial ecosystems. In Soil Microbial Association–Control of structure and function. Ed. Vancura V, Kune F. Academia Praha pp.339-405.

Thom C and Raper KB (1945) A manual of Aspergilli. Williams and Wilkinson Co. Baltimore 373 p.

Tiwari VK and Rai B (1977) Effect of soil burning on microfungi. Plant and soil 47 (3): 693-697.

Upadyay RS and Rai B (1979) Ecological survey of Indian soil fungi with special reference to Aspergilli, Penicilli and *Trichoderma* Rev. Ecol. Biol. Sol. 16 (1): 39-49.

Waksman SA (1922) A method for counting number of fungi in soil. J. Bot. 7: 339-341.

Wallwork JA (1976) The distribution and diversity of soil fauna. Academic Press, London.

Warcup JH (1950) The soil plate method for isolation of fungi from soil. Nature, London, 166: 117-118.

Wardle DA and Lavelle P (1997) Linkage between soil biota, plant litter quality and decomposition. In: Plant litter, quality and decomposition Ed. Cadisch G and Giller KE, CAB International, Wallingford pp 107-124.

Zou X, Zucca CP, Robert B, William H, McDowell (1995) Long term influence of deforestation on tree species composition and litter dynamics of a tropical rain forest in Puerto Rico. Forest Ecology and Management. 78: 147-157.

Chapter 12

Ichthyic Fauna Conservation of Hirakud Reservoir

M.M. *Prasad* [1*], *J.K. Bandyopadhyay* [1] *and Prem Kumar* [2]

[1]*Central Institute of Fisheries Technology*
Burla Research Centre, Burla – 768 017, Orissa
[2]*Experimental Fish Farm and Field Center, NRC CWF,*
Collecterate Road, Champawat, Uttaranchal – 262 523

ABSTRACT

Hirakud Reservoir on Mahanadi River System is largest reservoir in India. Located in between 20°31N, 85°10E Latitude and 21°45N 83° E Longitude, this reservoir encompasses vast area of 75000 hectares and is source of great ichthyic diversity. However, in the absence of inland fisheries legislation or the impediments associated with implementing the one, paved a way for fishing activities employing eco-destructive fishing gears. The use of low mesh sized fishing gears to harvest forage fishes and juveniles of commercially important fishes has reduced the original fish stocks resulting in sharp decline in catches from 750 tons in 1978-79 to 151 tons in 2004-05. This also resulted in significant decrease in catches of major carps. Continuation of this trend can have dangerous impact on the

---

* Corresponding Author: Dr. M.M. Prasad, Ph.D., ARS, Scientist in Charge; Ph: (O) 0663-2430503, (R) 0663-2432619, Fax: 0663-2430419; E-Mail: prasadmm@hotmail.com, prasadmothadaka@yahoo.com.

diversity of ichthyic fauna of Hirakud Reservoir. In this chapter the reasons for the present problems and suggestions to overcome the same are discussed.

***Keywords**: Conservation, Fish, Hirakud Reservoir.*

Introduction

Hirakud Reservoir on Mahanadi River System is the largest reservoir in India. Total area of the reservoir quoted by Jhingran (1983) and, Raghavachari and Rao (1984) is 74,592 hectares and 71,400 hectares respectively. Geographically, Hirakud is situated in between 20° 31'N, 85° 10'E latitude and 21° 45'N, 83° E longitude. The physical features of Hirakud Reservoir are presented in Table 12.1. In Mahanadi River System, Ib River is the largest tributary followed by Sheonath, Oang and Tel. Details of Hirakud Reservoir are shown in Figure 12.1. During construction of the dam thousands of families lost their shelters and agricultural lands. These disturbed families are settled around the reservoir in groups of 100-150 families. Most of the population is landless labour thereby they adopt fisheries as main profession as there is no alternative for employment. The average yield was at its peak during late seventies to early eighties when it was recorded to be 11 kg per hectare per year, which gradually reduced to 3.1 kg per hectare in 2000-2001. The declining trend in fish catches in preceding decade is depicted in Table 12.2. This happened due to indiscriminate killing of brooders and juveniles of commercially important fishes using restricted gears. Presently, many resident species show a decline in catches that may lead to serious problems to the biodiversity of the largest lacustrine ecosystem of India. The fisheries resources of Hirakud Reservoir, their production potential, periodic changes in biodiversity, causes of low productivity and suggestions for future strategies have been discussed hereunder.

Materials and Methods

It is known fact that the fish catches in Hirakud Reservoir are dwindling. This needs scientific analyses based on the data available and also assessment of the factors affecting the same. In the present study we have carried out extensive survey work on fish catch availability, gears employed in different fishing hamlets adjoining Hirakud Reservoir and post harvest fish utilization scenario and its

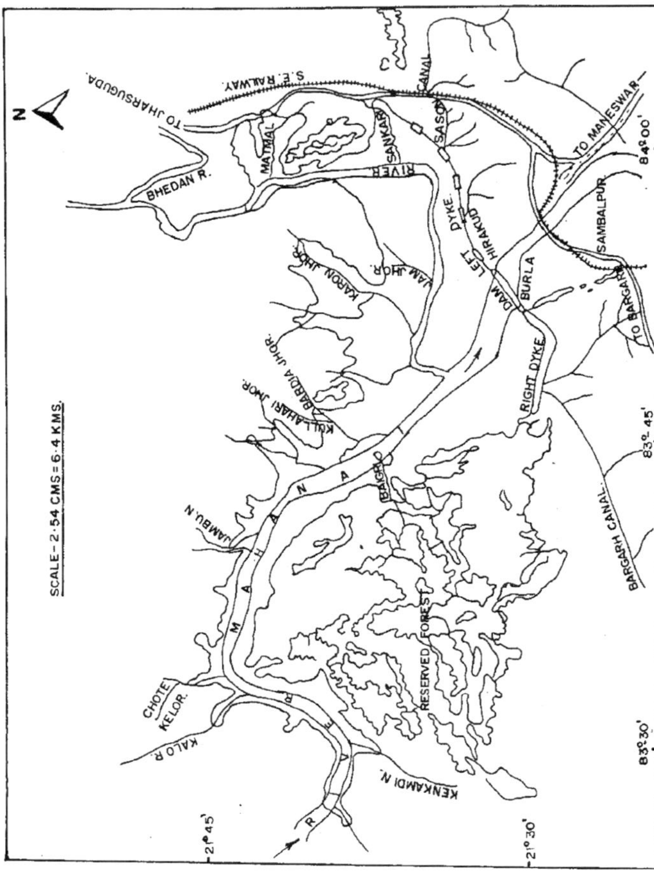

Figure 12.1: Map of Hirakud Reservoir

impact on the ichthyic diversity of the Hirakud Reservoir itself. Different factors influencing the ichthyic fauna diversity were identified through survey work and based on data collected both during experimental fishing and through the data generated from the Burla Research Centre of Central Institute of Fisheries Technology.

Table 12.1: Physical Features of the Hirakud Reservoir

1.	Maximum depth (DRL)	179.026 m
2.	Maximum depth (FRL)	192.017 m
3.	Total storage capacity	6.6 m/a/ft
4.	Live storage capacity	4.72 m/a/ft
5.	Catchment area	83395 sq.km
6.	Shore line	643.6 km
7.	Water spread area	460.8 sq.km

Table 12.2: Fisheries Production in Hirakud Reservoir

Year	*Production in M.T.*	*Decrease in Production (MT)*	*Decrease in Per cent Production*
1995–1996	358.6	–	–
1996–1997	267.3	91.3	25.46
1997–1998	260.2	98.4	27.44
1998–1999	274.8	83.8	23.36
1999–2000	259.24	99.2	27.33
2000–2001	235.00	123.6	34.47
2001–2002	186.31	172.29	48.00
2002–2003	199.5	159.1	44.36
2003–2004	163.78	194.82	54.32
2004–2005	151.54	207.6	57.74

Results and Discussion

Fisheries Resources

Job *et al.* (1955) surveyed the fisheries resources of "Mahanadi River System" and found 103 species of which 24 were important commercially. Khan *et al.* (1992) described 43 species in Hirakud Reservoir out of which 18 are commercially important. The ichthyic fauna of the reservoir with local names are presented in Table 12.3. The data in Table 12.4 indicates the commercially important species and their per cent contribution of the total catches of the reservoir. The studies carried out in recent years show that catches comprise mainly of cat fishes like *Eutropichthys vacha, Wallago attu, Mystus singhala, M. aor, Rita chrysea, Silonia silondia* etc., accounting for 40-50 per cent, followed by carps *viz. Labeo fimbriatus, L. calbasu, L. rohita, Cirrhinus mrigala, Catla catla etc.* for 15-18 per cent and rest are miscellaneous fishes like *Gudusia chapra, Rohtee cotio, Chela bacaila and Ailea coilea* (Anon, 2001).

Table 12.3: Fish Fauna of Reservoir

Sl.No.	*Name of the Fish*		*Local Name*
1.	*Catla catla*	(Hamilton)	Bakur
2.	*Labeo fimbriatus*	(Bloch)	Pursi/Berang
3.	*Labeo calbasu*	(Hamilton)	Kalanchi
4.	*Labeo rohita*	(Hamilton)	Rohu
5.	*Labeo gonius*	(Hamilton)	Nunia
6.	*Labeo bata*	(Hamilton)	Bhangen
7.	*Cirrhina mrigala*	(Hamilton)	Mrikali
8.	*Cirrhina reba*	(Hamilton)	Pohura/Dumala
9.	*Barbus tor*	(Hamilton)	Kusra
10.	*Puntius serrana*	(Hamilton)	Sarna
11.	*Mystus seenghala*	(Hamilton)	Aadi/Singhari
12.	*Mystus aor*	(Hamilton)	Katrang
13.	*Mystus gulio*	(Hamilton)	Tingra
14.	*Silonia silondia*	(Hamilton)	Gajra
15.	*Wallago attu*	(Schneider)	Balia
16.	*Pangasius pangasius*	(Hamilton)	Jalang
17.	*Rita chrysea*	(Day)	Kukia
18.	*Eutropichthys vacha*	(Hamilton)	Vacha
19.	*Bagarius bagarius*	(Hamilton)	Buthia
20	*Notopterus notopterus*	(Pallas)	Fali

Contd...

Table 12.3–Contd...

Sl.No.	Name of the Fish		Local Name
21.	*Bitioterys ogutaka*	(Hamilton)	Ogutak
22.	*Gudusia chapra*	(Hamilton)	Phufur
23.	*Rohitee cotio*	(Day)	Chilanti
24.	*Pama pama (Sciaenid)*	(Hamilton)	Patharmundi
25.	*Glossogobius guiris*	(Hamilton)	Ghesra
26.	*Rhinomugil corsula*	(Hamilton)	Koinga
27.	*Xenontodon cancilla*	(Hamilton)	Gourchena
28.	*Chela chela*	(Day)	Jedda
29.	*Chela bacailla*		
30.	*Ailea coilea*	(Hamilton)	Bauspati
31.	*Ambassis nama*	(Hamilton)	Bonju
32.	*Ambassis sophore*		Patponia
33.	*Puntius sophore*	(Hamilton)	
34.	*Puntius tocto*		Kutri
35.	*Puntius cola*	(Hora)	
36.	*Puntius dorsalis*		Kutri
37.	*Mastocembelus armatus*	(Lacepede)	
38.	*Mastocembelus pancalus*	(Hamilton)	Bayarri
39.	*Channa striatus*	(Hamilton)	
40.	*Channa marulius*	(Hamilton)	Soul
41.	*Channa punctatus*	(Hamilton)	Khapsi
42.	*Ompok bimaculatus*	(Bloch)	Banju
43.	*Machrobrachium malcomsonii*		Chingri

Production and Potential

According to Raghavachari and Rao (1984) and Varghese *et al.* (1992) average production of Hirakud Reservoir was around 750 tonnes per year during 1978 to 1981, which accounts for the productivity to be 11 kg per hectare. According to the State Fisheries Department the production reduced to 380 tonnes in 1991-92, 426 tonnes in 1993-94, 267 tonnes in 1996-97 and 235 tonnes in 2000-2001. The data of the last ten years shows a visible declining trend in fish catches (Table 12.2). The potential productivity of the reservoir has been estimated to be 15 kg per hectare (Khan *et al*, 1992). On the basis of this estimate, total production of the reservoir can be enhanced up to 1071 tonnes. The present production of the reservoir is only 14.15 per cent of the potential.

Table 12.4: Percentage Representation of Major Fishery of the Reservoir

Sl. No.	Name of the Fish	Percentage in Total Landings
1.	*M. seenghala*	14.66
2.	*S. silondia*	12.77
3.	*C. catla*	10.26
4.	*R. cotio*	7.4
5.	*W. attu*	7.3
6.	*L. calbasu*	7.06
7.	*L. fimbriatus*	6.77
8.	*G. chapra*	5.57
9.	*L. rohita*	5.13
10.	*N. chitala*	4.48
11.	*E. vacha*	3.78
12.	*L. bata*	3.12
13.	*R. chrysea*	2.35
14.	*C. mrigala*	1.49
15.	*Chela* sp.	0.65
16.	*Sciaenid (Pama pama)*	0.62
17.	*Barbus tor*	0.44
18.	*Charra* spp.	0.42

Reasons for Low Productivity

The reasons for low productivity of the reservoir are numerous but the principal ones have been described hereunder:

1. Use of Destructive Fishing Gears
2. Fishing Activities in Closed Season
3. Impediments in Implementation of Fisheries Legislation
4. Sedimentation of Reservoir

5. Obstructions in Fishing
6. Continuous enhancement in fishing efforts in Unit area
7. Climatic factors
8. Insufficient documentation
9. Biological factors.

Use of Destructive Fishing Gears

Stick-held Drag net, Shore seine, Gill net, Cast net, Hook and line and traps are the main gears employed in the reservoir. Greed for more catches has resulted in reduction of mesh size of fishing gears to such an extent that they even use mosquito nets for fishing in shallow waters. The catches of mosquito net are mainly weed fishes and juveniles of commercially important fishes. The juveniles of carps are either sold to the fish farmers after segregation or entire catches are smoked and sold in local markets. State Fisheries Department has estimated this by-catch to be about 35 per cent of total production of the reservoir. Indiscriminate use of these gears stops the scope for replenishment of the fish stock since young ones are caught in huge quantities regularly. Interestingly, fishermen are aware of the consequences of using these gears but the feeling of "*If I don't catch some other will..*" encourage them to destroy the nation's treasure. The Disco net (locally known as 'Dhulungi Jal') operation in Hirakud Reservoir and its negative impact is reported recently (Anon, 2005).

Fishing Activities in Closed Season

Although fishing is prohibited in Hirakud Reservoir between 15 June to 15 August every year which is breeding season for fish, the fishing activities are continued even in this period (Anon, 2005). The brooders of carps migrate to the upper reaches of the reservoir for breeding. Fishermen who are aware of their movements block their way by gill nets and kill them before breeding takes place. The few lucky brooders are able to reach the breeding ground escaping this death trap, but their juveniles cannot escape from their mosquito net.

Impediments in Implementation of Fisheries Legislation

Use of restricted gear and fishing in the closed season is the result of lack in fisheries legislation for inland waters in the state.

There are reports that the State Fisheries Department failed to restrict the fishing activities during breeding season on many occasions because of this lacuna. However, very recently the Government of Orissa enacted a law to protect the reservoir from illegal fishing (Anon, 2004). According to the same;

"In exercise of the powers conferred under clause-8 of the "State Reservoir Policy, Orissa", the state govt. do hereby declare the officers in the rank of Fishery Extension Officer/Inspector of Fisheries/Deputy Superintendent of Fisheries/Junior Fisheries Officer/Assistant Director of Fisheries as Authorized Officer for enforcement of the conservation measures as defined under clause–7 (7.1 to 7.10) of the "State Reservoir Fishery Policy" to perform the functions and exercise the powers under the said clauses of the policy within their respective jurisdiction in the State of Orissa and they shall exercise following powers.

1. To inspect any fish, fishery products, fishing crafts, gears, plants, machineries, hatcheries, fish landing platforms/ sheds, jetties etc.
2. During the course of inspection if any Authorized Officer detects unauthorized fishing from a reservoir, the Authorized Officer may take charge of the stock to caught and dispose off the stock in the open market at the prevailing price. The Authorized Officer shall thereafter lodge an FIR in the nearest police station.

Sedimentation of Reservoir

Anon, (1986) reported the sedimentation rate of the Hirakud reservoir is 23.32 Mm^3 per year and 62 per cent of sedimentation is deposited in the upper 38 per cent of the reservoir depth. Fish breeding grounds are generally located in the upper reaches of the reservoir; this sedimentation is reducing the capacity of the breeding grounds. High concentration of suspended solids in the water cause low hatching rate of eggs and high mortality of the spawn and fry due to oxygen depletion.

Obstructions in Fishing

Khan *et al.* (1992) stated that 240 villages were submerged during the construction of the dam across the river Mahanadi that resulted in the formation of Hirakud Reservoir. The reservoir bed is littered with stones, boulders, tree stumps and submerged villages etc. that account for 40-45 per cent of the total area of the reservoir. As per the estimate of the state fisheries authorities, the actual fishable area is only about 17 per cent of the total water spread. These submerged trees are causing destruction of the nets. This is another indirect reason of lower productivity of the reservoir.

Continuous Enhancement in Fishing Efforts in Unit Area

More often the fact that is not taken into consideration is, there is no change in the area of reservoir in general and fishing grounds in particular. Rather, in due course of time for various reasons the size of the reservoir or the fishing grounds decrease. Contrary the fishing efforts increase continuously due to increase in population, which is also a major factor in decrease in productivity.

Climatic Factors

Impact of climatic factors such as drought was never been assessed on the productivity of the reservoir. Variations in water level in different seasons especially during drought can result in changes in breeding grounds that can adversely affect the productivity of the reservoir at a later stage.

Insufficient Documentation

The present documentation on production did not take into consideration certain prohibited sectors of reservoir, which are highly productive. Also the data obtained is limited to licensed fishermen only. This does not represent poached, illegal or unauthorized fishing. There is need for development of mathematical models and simulation studies to predict trends in the production. Time taken in documentation of data and statistical analyses are also important factors in projecting the accurate productivity levels of the reservoir.

Biological Factors

The pressure of relentless harvest can have significant impact on the population, especially commercially important fishes. This can result in disappearance of species and appearance of another species. One such example is freshwater prawn *Macrobrachium malcolmsonii*, which is reported to occur in 1950s in Mahanadi River System, again it is seen in 1970s and after a gap of two and half decades its occurrence seen these days. Migratory nature of fish, changes in habitats which are endemic to certain varieties of fish, impact of agricultural pollution such as pesticides, insecticides, industrial effluents seepage into reservoir and its their impact on food web of the fishes, predators mainly birds are not taken into consideration when the productivity levels of reservoir are assessed. The impact of birds on the fish catch world over in marine sector is well documented (Michael Brooke, 2004). The consumption of fish by birds is staggering (Table 12.5) However, similar kind of data is not available in Indian context both in marine and inland fisheries sectors that included reservoirs such as Hirakud Reservoir.

Table 12.5: Consumption of Fish by Seabirds

Sl.No.	*Principal Groups of Birds*	*Consumption of Fish (Million Tons)*
1.	Petrels; Albatross (Procellariforms)	24.1
2.	Penguins	23.6
3.	Auks	11.2
4.	Terns and Skinners	33.9
5.	Gulls and Skuas	3.7
6.	Cormorants	1.5
7.	Boobies, Garnets, Tropic bird and Frigate birds	1.2
8.	Pelicans	0.5

The totals represent base estimates.

Suggestions for Increasing Productivity and Conservation of Biodiversity

Taking in to account the low productivity of the reservoir some suggestions are given hereunder which may enhance not only productivity but at the same time conserve the biodiversity also.

1. The state government must enact Fisheries Legislation for inland waters. The ban on indiscriminate use of fishing gears and fishing activities during breeding season must be imposed strictly.
2. The fishermen community around Hirakud reservoir is very poor and there is no alternative source of income during rainy season for which they have to continue fishing activities for their livelihood. Therefore, efforts are needed to make them aware of various domestic level

enterprises such as Aquaculture, Bee keeping, Silk worm culture, Poultry production and Mushroom culture etc. as alternative sources of income during lean fishing season.

3. Re-stocking of the reservoir should be done with fingerlings of commercially important fishes. CIFT is taking an initiative in this direction under special component plan.
4. Obstruction of fishing *i.e.* submerged tree trunks can be marked with buoys and beacons to avoid unnecessary destruction of nets.

Conclusion

Employment of fishing gears of extremely small mesh size and fishing activities during breeding season has led to overexploitation of the resources. This has endangered the existence of the bio-diversity of India's largest lacustrine ecosystem that has to be repaired by imposing ban on eco-destructive gears and restocking of the reservoir. In the conservation of biodiversity, the need of the hour is coordinated efforts between research, educational institutions, different governmental and non-governmental agencies, in disseminating technologies and also in developing pragmatic awareness programs that could reach masses.

Acknowledgement

Authors are thankful to Dr. Devadasan, Director, CIFT, Cochin, for permission to present and publish this paper in National Symposium on "Environmental Biotechnology and Biodiversity Conservation".

References

Anon (1986) Third report on Sedimentation survey in Hirakud Reservoir. Hirakud Research Station, Department of Water Resources, Govt. of Orissa.

Anon (1997) State Fisheries Statistics, Directorate of Fisheries, Govt. of Orissa.

Anon (2001) Annual Report. Central Institute of Fisheries Technology. Matsyapuri P.O. Kochi 682 029.

Anon (2004) Government of Orissa: Fisheries and A.R.D. Department (Notification No-1-Fy-Sch.-2/20039905/FARD, D: 29.05.2004, issued By the Order of Governor; A.K. Tripathy, Commissioner cum Secretary to Government of Orissa).

Anon (2005) Orissa News Letter. Fishing Chimes. 25 (9):22-23.

Khan, AA, Dawson, P, Rao, JSR and Varghese, MD (1992)Fishing in Impounded Waters,. A Case Study of Hirakud Reservoir–Orissa. CIFT Publication.

Job, TJ, David, A and Das, KM (1955) Fish and Fisheries of Mahanadi in relation to Hirakud Dam. Ind. J., Fish. 2.1.

Jhingran, VG (1983) Fish and Fisheries of India. Hindustan Publishing Corpn. Delhi. 170-199.

Michael Brooke (2004). "The food consumption of world's seabirds" (*Biology, Letters Proceedings February 2004*). World Fishing 53 (3):1.

Raghavachari, M. and Rao, S (1984) Inland Fisheries Resources in India. Concept Publishing Company New Delhi. 213-250.

Varghese, MD, Nair, AKK, George, VC and Khan, AA (1981) Fish. Technol. 18.17-23.

Chapter 13

Quantitative Analysis of Herbaceous Vegetation in a Dry Deciduous Forest of Sambalpur, Orissa

Geeta Rani Pattanaik and S.L.Swain*

P.G. Department of Botany, G.M. College (Autonomous)
Sambalpur – 768 004

ABSTRACT

The study deals with the quantitative analysis of herbaceous vegetation in a dry deciduous forest of Sambalpur, Orissa. The study site named Motijharan Reserve Forest located in the heart of Sambalpur Town at 21°- 30' N latitude and 84° – 30' E longitude. The analysis of herbaceous vegetation revealed the presence of a total of 35 species in the study area. Species like *Cynodon dactylon, Evolvulus alsenoides, Sida cordifolia* and *Tridax procumbens* were the common herbaceous species in all the forest sites of the study area. In the present study species diversity ($\overline{H}$) exhibited maximum value during rainy season and minimum during summer. The relationship between mean value of diversity and dominance was observed to be inverse (r= -0.98). Diversity was observed to be least in deforested barren site ($\overline{H}$ = -1.858).

Keywords: *Herbaceous vegetation, Importance value index, Species diversity (H), Concentration of dominance (Cd), Deforested barren site.*

---

* Corresponding Author.

Introduction

Herbaceous ground vegetation, though constitutes a small proportion of the total biomass in the forest ecosystem, plays a very important role in classification and ecological characterization of forest (Ovington, 1959; Ovington and Madgwick, 1959; Whittaker, 1966). The importance of herbaceous ground vegetation in forest ecosystem has been stressed due to its high nutrient content and rapid turnover rate (Edwards and Grubb, 1977; Pandey and Singh, 1981). As per Hughes (1971), ground vegetation contributed around 20 per cent to the total above ground production in mixed deciduous woodland. In a study of the dry matter dynamics of a pine forest of Central Himalayas, Chaturvedi and Singh (1987) reported that herbaceous vegetation of the forest floor contributes around 15 per cent to the total net primary production. Because of the higher turnover of herbaceous litter, herbaceous stratum is also expected to have a major influence on the nutrient dynamics in the forest ecosystem. Therefore, in order to understand the functioning of the forest ecosystem with respect to energy flow and nutrient cycling, it is quite essential to get information about the herbaceous vegetation.

Material and Methods

Study Site

The study site, a dry deciduous natural forest on a hillock named Brooks-hill, locally named as Motijharan hill, located in the heart of Sambalpur Town at 21°–30' N latitude and 84°–30° E longitude. It has an area of 63.585 hectares. The highest point of the hill is 802 feet. The foot hill portions are plain to gentle sloping. The importance of Motijharan hill forest lies with the fact that it is one of the natural dry deciduous forest in the centre of the town. There is a Television Station and Television Tower on the top of the hill. The main attraction of the study site is the presence of a "Deer Park" under the care of "Wild Life Management".

The climate of the area is seasonal tropical in nature with three distinct season (Summer, rainy and winter). Annual rainfall of the area is 1272 mm, out of which 87 per cent falls during rainy season (June–September). The mean air temperature of the area varies from as low as 13.06°C in December to as high as 45°C in May. The relative humidity of the area ranges from 25 per cent during April to 89 per cent during July-August.

Soil is usually shallow to deep sandy loam containing some stony nodules at the foot hill and slope region. The soil of the area is red lateritic type. The colour is due to leaching of soluble base during rainy season and oxidation of Iron and Aluminum compounds during summer season. The foot hill portion has good depth of alluvial soil with small pebbles. Some of the soil physico-chemical characteristics are mentioned inTable 13.1, considering the heterogeneity of the area with respect to vegetation growth four sites were selected for the study.

Table 13.1: Soil Physico-chemical Characteristics of the Study Area

Sl.No.	*Parameters*	
1.	Soil Texture (per cent)	
	Clay	7–11 per cent
	Silt	9–24 per cent
	Sand	84–65 per cent
2.	Bulk density (g/cm^3)	1.17–1.57
3.	Water holding capacity (per cent)	27.2–57.5
4.	pH	5.8–6.5
5.	Organic Carbon	0.566–1.998
6.	Total nitrogen	0.472–1.7
7.	C:N ration	0.720–1.14

Quantitative analysis of the herbaceous vegetation was carried out at three different seasons on four different sites of the study area. Ten quadrants, each of 0.5 m × 0.5 m size were laid down at random at each of the region. The analytical measures, such as relative frequency, relative density and relative abundance were determined for all the comprising species (Misra, 1968). The importance value index (IVI) of the ground flora species were calculated by summation of relative value of frequency, density and abundance (Curtis, 1959). Species diversity ($\overline{H}$) was calculated following the formula of Shannon and Wiener (1963). Concentration of dominance (Cd) was calculated following Simpson (1949).

$\overline{H}$: $\Sigma[^{ni}/_N \text{ Log}_e \, ^{ni}/_N]$

Cd: $\Sigma(^{ni}/_N)^2$

where,

$\overline{H}$: Diversity index

Cd: Concentration of dominance

ni: IVI of individual species

N: IVI of all species

Results

Analysis of herbaceous vegetation along with their IVI value during three different seasons and at four different sites of the study area are presented in Tables 13.2–13.5. The study revealed the presence of all total of 35 species. *Cynodon dactylon, Evolvulus alsenoides. Sida cordifolia, Tridax procumbens* were common to all sites. Comparison of IVI of different herbaceous species in four different sites of the study area revealed that in site I all total of 27, 17, 7 species were recorded during rainy, winter and summer seasons respectively. *Ocimum canum* was observed to be the dominant herb exhibiting maximum IVI during rainy (IVI = 0.253), during winter season (IVI = 0.47) and during summer (IVI = 0.88) because of its greater mean basal area.

Among the 14 species distributed in site II during rainy and winter IVI value was observed to be maximum for *Ocimum basilium* (IVI = 04.76 and 0.372 respectively). However during summer *Ocimum canum* exhibited maximum IVI (0.363) value.

In the site III, all total of 21 species were recorded during rainy season of which *Ocimum canum* was the dominant species IVI = 0.32. During winter season *Atylosia scarabaeoides* exhibited the maximum IVI (0.46). However during *summer, Sida cordifolia* exhibited the highest IVI value (0.82). *Vandellia crustacea* was the species with least IVI value (0.08) in this site.

Among the 7 species distributed in deforested barren site, maximum IVI value was recorded for *Cynodon dactylon*. During rainy, winter and summer season IVI = 0.978, 0.88, 0.655 as it was frequently distributed species. In this site *Oldenlandia corymbosa* was the species with least IVI value (IVI = 0.291).

Table 13.2: Analysis of Herbaceous Vegetation and their IVI During Three Different Seasons Towards Site–I

Name of the Species	IVI Value		
	Rainy Season	Winter Season	Summer Season
Ageratum conyzoides L.	0.086		
Andrographis paniculata Ness.	0.102	0.16	0.25
Atylosia scarabaeoides Benth.	0.171	0.16	0.39
Belpharis molluginifolia Pers.	0.096		
Cassia tora L.	0.159		
Cleome Viscosa L.	0.111	0.10	
Cynodon dactylon Pers.	0.109	0.22	0.30
Desmodium troiflorum DC.	0.666		
Euphorbia hirta L.	0.099	0.07	
Evolvulus alsenoides L.	0.135	0.10	
Heliotropius indicum L.	0.081	0.20	0.25
Heteropogon contortus Roem	0.066		
Indigofera pulchella Roxb.	0.109		
Justicia simplex D.Don.	0.098		
Mimosa pudica L.	0.095	0.16	
Ocimum basilicum L.	0.123		
Ocimum canum Sims.	0.253	0.47	0.88
Phyllanthus niruri L.	0.083	0.10	
Panicum caesium Nees.	0.056	0.12	
Panicum indicum L.	0.066	0.14	
Sida cordifolia L.	0.122	0.22	0.56
Spermacace hispida L.	0.095		
Tephrosia purpurea Pers.	0.142	0.12	0.25
Tridax procumbens L.	0.099	0.245	
Urena lobata L.	0.066		
Vandellia crustacea Benth	0.096	0.10	
Vernonia cineria L.	0.066	0.16	

Table 13.3: Analysis of Herbaceous Vegetation and their IVI During Three Different Seasons Towards Site–II

Name of the Species	IVI Value		
	Rainy Season	Winter Season	Summer Season
Aristida adscenscinis L.	0.421	0.274	0.247
Cleome viscosa L.	0.286		
Croton sparsiflorus Morung.	0.12		0.322
Celosia argentea L.	0.183		
Cynodon dactylon Pers.	0.147	0.244	–
Euphorbia hirta L.	0.14	–	–
Evolvulus alsenoides L.	0.161	–	–
Ocimum basilicum L.	0.476	0.2372	–
Ocimum canum Sims	–	–	0.363
Sida cordifolia L.	0.377	–	–
Sida veronicaefolia L.	0.24	0.287	0.299
Scoparia dulcis L.	0.271	–	–
Tridax procumbens L.	0.219	0.202	–
Urena lobata L.	–	0.368	0.263

Table 13.4: Analysis of Herbaceous Vegetation and their IVI During Three Different Seasons Towards Site–III

Name of the Species	IVI Value		
	Rainy Season	Winter Season	Summer Season
Andrographis paniculata Nees.	0.08	0.18	0.34
Atylosia scara baeoides Benth.	0.26	0.46	0.54
Cassia tora L.	0.13	–	–
Cynodon dactylon Pers.	0.11	0.18	0.40
Cyperus rotundus L.	0.15	–	–
Desmodium triflorum D.C.	0.13	–	–
Euporbia hirta L.	0.11	–	–
Euphorbia alsenoides L.	0.11	0.15	–
Indigofera pulchella Roxb.	0.11	–	–

Contd...

Table 13.4–Contd...

Name of the Species	IVI Value		
	Rainy Season	Winter Season	Summer Season
Leucas aspera Spreng.	0.15	–	–
Mimosa pudica L.	–	0.24	–
Ocimum basilicum L.	0.13	0.15	0.34
Ocimum canum Sims.	0.32	0.34	0.47
Panicum indicum L.	0.11	–	–
Phyllanthus niruri L.	0.11	0.15	–
Sida cordifolia L.	0.15	0.29	0.82
Spermacoce hispida L.	0.15	0.15	–
Tephrosia purpurea Pers.	0.11	0.18	–
Tridax procumbens L.	0.13	0.27	–
Urena lobata L.	0.11	0.16	–
Vandellia crustacean Benth	0.08	–	–

Table 13.5: Analysis of Herbaceous Vegetation and their IVI During Three Different Seasons Towards Site–IV (Deforested Barren Site)

Name of the Species	IVI Value		
	Rainy Season	Winter Season	Summer Season
Cynodon dactylon Pers.	0.978	0.88	0.655
Evolvulus alsenoides L.	0.426	0.401	0.39
Oldenlandia corymbosa L.	0.291	–	–
Scoparia dulcis L.	0.31	–	–
Sida acuta Burm.	0.595	0.515	0.315
Sida cordifolia L.	0.434	0.388	–
Tridax procumbens L.	0.538	0.413	0.307

Table 13.6 represents the number of species, diversity ($\overline{H}$) and dominance (Cd) indices in four different sites of the study area during different seasons. In all the sites, maximum number of species were recorded during rainy season where as, during summer the number of species were the least. Value of herbaceous species diversity ($\overline{H}$)

showed a range of 1.858 in site IV (deforested barren site) to 3.18 in site I. In all the sites, diversity was calculated to be maximum during rainy season and minimum during summer season.

Herbaceous dominance index (Cd) varied from 0.041 in site-I during rainy to 0.22 in site IV during summer season. In all the sites, the dominance index was maximum during summer and minimum during rainy season. The relationship between mean value of diversity and dominances was observed to be inverse (r = -0.98).

Table 13.6: Annual Variation of Number of Species (S) Diversity Index ($\overline{H}$) and Concentration of Dominance (Cd) of Herbaceous Species in Different Sites of the Study Area

Sites	*Rainy*	*Winter*	*Summer*	*Mean*
		Site–I		
No. of species	27	17	7	17
$\overline{H}$	-3.18	-2.699	-1.777	-2.552
Cd	0.0405	0.0735	0.1413	0.085
		Site–II		
No. of species	14	10	8	10
$\overline{H}$	-2.669	-2.257	-1.809	-2.245
Cd	0.087	0.089	0.098	0.091
		Site–III		
No. of species	20	13	6	13
$\overline{H}$	-2.896	-2.598	-1.706	-2.4
Cd	0.0571	0.886	0.1775	0.108
		Site–IV		
No. of species	7	5	4	5
$\overline{H}$	-1.549	-1.549	-1.331	-1.579
Cd	0.164	0.223	0.275	0.22

Correlation coefficient between 'r' value

1. Mean diversity index and mean dominance index–0.98
2. Herbaceous species number and Herbaceous diversity 0.964.

Discussion

Analysis of herbaceous ground vegetation revealed that species like *Cynodon dactylon, Evolvulus alsinoides, Sida cordifolia* and *Tridax procumbens* were the common herbaceous species during different seasons in all the four sites of the study area. Since herbaceous species generally have short life span and exhibit quick response to change of different environmental parameters like soil moisture, temperature, relative humidity shedding effects of tree canopy cover etc. there observed a change with respect to their dominance during different seasons in four selected study sites.

In the present study, species diversity ($\overline{H}$) exhibited maximum value during rainy season and minimum during summer. This may be due to the influence of favorable moisture and temperature during rainy season. The dominance index (Cd) was observed to be maximum during summer and minimum during rainy season. The relationship between the number of herbaceous species and the herbaceous diversity was observed to be positive (r = 0.96). Since diversity is a measure of species richness, perhaps qualitatively, species rich *Phanenophytic* vegetation on the site II did not allow more diversified herbaceous community to grow under it. However, a few selected herbaceous shade loving species in the site II can compete to grow under the thick canopy of the larger trees. Hence, rise in canopy cover may effect the herbaceous diversity on the forest floor through light intensity (Saxsena and Singh, 1982).

In the deforested barren site (site IV) there is no tree or shrub vegetation. It supports only a feeble development of herbaceous flora. Diversity was observed to be least in deforested barren site ($\overline{H}$ = 1.858) in comparison to other sites. Loss of tree vegetation and least herbaceous diversity in deforested barren site is due to substantial decline in clay particles and loss of organic carbon and nitrogen (Hajabbasi *et al.*, 1997; Sahani and Behera, 1998).

References

Chaturvedi OP and Singh JS (1987) Structure and function of Pine forest in Central Himalayas 1. Dry matter dynamics Annals Bot. 60:237-252.

Curtis JT (1959) The vegetation of Winconsin: an ordination of plant communities. Medison Winconsin Press, Madison, Winconsin.

Edwards PJ and Grubb PJ (1977) Studies of mineral cycling in a montane rain forest in New Guinea. 1. The distribution of organic matter in the vegetation and soil. J. Ecol. 65: 943-969.

Hajabbasi MA, Jalilian A and Karimzadeh HR (1997).Deforestation effects on soil physical and chemical properties, Lordegan, Iran. Plant and Soil 190: 301-308.

Huges MK (1971) Ground Vegetation, biocontent and net production in a deciduous Woodland. Oecologia (Berl) 7: 127-135.

Misra R (1968) Ecology Work Book. Oxford and IBH Publication Co.

Ovington JD (1959) The circulation of minerals in plantation of *Pinus sylvestris* L. Annals Bot 23: 229-239.

Ovington JD and Medgwick, MA. (1959) Distribution of Organic matter and Plant nutrient in a Plantation of Scot Pine. Forest Science 5: 344-355.

Pandey U and Singh JS (1981) A quantitative study of the forest floor, litter fall and nutrient return of Oak-conifer forest in Himalayas. II. Pattern of litter fall and nutrient return Oecol. 2(1): 83-99.

Sahani U and Behera N (1998) Impact of Deforestation on Physico-chemical characteristics of Tropical Soil. Indian J. Env. Ecoplan 1(1 and 2): 33-38.

Saxsena AK and Singh JS (1982) A Phytosociological analysis of woody species in forest communities of a part of Kumaun Himalaya. Vegetation, 50: 3-22.

Shannon CE and Wiener W (1963) The mathematical theory of communication P.117 Urbana: University of Illinois Press.

Simpson EH (1949) Measurement of diversity Nature. 163: 668.

Whittaker RH (1966) Forest dimensions and Production in the great smoky mountains. Ecology. 47: 103-121.

Chapter 14

Probable Solution of Selecting Medicinal Plants Having Controversial Identification in Ayurvedic Therapy

B.N. Mahapatra [1]* and B.P. Choudhury [2]

[1]*Reader and Head,*
Department of Dravyaguna
Gopabandhu Ayurved Mahavidyalaya, Puri – 752 002 (Orissa)
[2]*Former Professor and Head, P.G. Department of Botany,*
Utkal University, Bhubaneswar – 751 004 (Orissa)

ABSTRACT

Use of plants or plant based medicines dates back to the human civilization. Ayurvedic therapy is mainly connected with medicinal plants. About 125 numbers of medicinal plants in *Atharva Veda* and subsequently 1000,573 and 902 numbers of medicinal plants in *Charak Samhita, Sushruta Samhita* and *Astanga Hridayam* respectively have been identified as medicinal plants. However, the sanskrit names have been mainly used in those treatise. It has been found later that, some medicinal plants have poly-botanical equivalents. For example Brahmi has been

* Mobile: 94373 30455; E-mail: bn_mohapatra@rediffmail.com.

used as *Bacopa monnieri* and *Centella asiatica;* Bhringaraj as *Eclipta prostrata, Wedelia chinensis, Limnophila rugosa;* Shankhapuspi as *Canscora decussata, Evolvulus alsinoides, Convolvulus pluricaulis, Clitoria ternatea;* Jeevanti as *Leptadenia reticulata, Holostemma annulare;* Pashanabheda as *Rotula aquatica, Aerva lanata, Kalanchoe pinnata, Polycarpea aurea* and Shalaparni as *Desmodium gangeticum, D. Polycarpum, Flemingia stricta.* This has created a lot of controversy and confusion in using the medicinal plants having a particular active principle for curing specific disease. At present medicinal plants have been scientifically identified by the help of Botanical keys and characters and have been attributed to binomial nomenclature in pursuance with the Principles and Articles of the Code (ICBN). Therefore, it is necessary to select a scientifically identified medicinal plant having specific active principle for the ancient sanskrit name of medicinal plant to be used in Ayurvedic therapy. Thereby the efficacy of this ancient therapy will gain its popularity and acceptability.

Keywords: *Ayurvedic therapy, Controversial identification, Medicinal plants.*

Introduction

Use of plants or plant based medicines dates back to the human civilization. Ayurvedic therapy, one of the ancient methods of treatment, remains as a land mark for curing various ailments. It is an off shoot of ethno-medicinal practice adopted by various ethnic groups. In course of time, it has undergone many modifications and has been adopted by the personnels which is indicated in the ancient practices. As a matter of fact, there is mention of about 125,1000,573 and 902 numbers of medicinal plants in Atharva Veda, Charak Samhita, Sushruta Samhita and Astanga Hridayam respectively.

Subsequently, the number of medicinal plants has considerably increased which ultimately attracted several foreigners to this land. The early Dutch, Portuguese and Britishers were greatly influenced by the healing properties of these plants. Many of these foreigners and some Indian writers brought out a large number of publications relating to the medicinal plants of this country. Some of the important publications in this context include *OS Colloquius* (Garciad, Orta, 1563), *Hortus Malabaricus* (Rheede, 1678-1703), *Hortus Bengalensis* (Roxburgh, 1814), *Flora Indica* (Roxburgh, 1832), *Flora of British India*

(Hooker,1872-1897) *Dictionary of economic products* (Watt, 1887-1893). The most comprehensive works on medicinal plants of this country were complied by Nadkarni (1976), Kirtikar and Basu (1918), Chopra *et al.* (1956), Satyavati *et al.* (1976-1987) Pullaiah (2002) and a few others.

In all the ancient treatise, sanskrit names/synonyms of medicinal plants have been cited there in. However, due to scanty phytography of medicinal plants there is often controversy in those Vedic medicinal plants. Out of 15,000 species of angiospermic plants in India, around 5000 species are having known medicinal properties. This increased number of medicinal plants, includes a number of substitutes and thereby there is gradual increase of controversial medicinal plants.

All the major systems of indigenous medicines–Ayurveda, Unani, Siddha, Homeopathy, Amchi, Tibbi etc. are largely based on plants only. In fact, over 120 pharmaceutical products currently in use are plant derived and more than 75 per cent of them were discovered by the clues obtained from traditional healers. To quote some exmples 'Quinine' an alkaloid isolated from the bark of Cinchona tree (*Chincona calisaya* and *C. officinalis*), originally a native of South America, is the oldest of all anti-malarial drug. Similarly, d-TUBOCURARINE, an alkaloid obtained from the plant *Chondrodendron tomentosum* indigenous to Western Amazon region, and plants of *Strychnos* species (mainly *Strychnos toxifera*) from Eastern Amazon region of South America has been used as skeletal muscle relaxant. Curare was mainly used as an arrow poison by the South American aborigines, although the mechanism of action of Curare was studied in 1856 and the pure alkaloid was isolated only in 1935. The bitter principle 'Chirata' of Indian Pharmacopoeia (I.P) used to stimulate appetite is obtained from *Swertia chirata*, which was a common house hold remedy in India. Among the digestive enzymes 'Papain' is obtained from vegetable source (*Carica papaya*) having proteolytic enzyme. Some Indian food contains enough carminatives in various spices and condiments such as cardamom seeds (*Elettaria cardamomum*), ginger (*Zingiber officinale*), coriander (*Coriandrum sativum*), asafoetida (*Ferula narthex*), cinnamon bark (*Cinnamomum zeylanicum*) and clove (*Syzygium aromaticum*). Therapeutically, volatile oil of these substances are used in the form of tincture.

Ayurvedic therapy gained much popularity after independence of India, although it faced critical competition with allopathic therapy adopted by English people. Allopathic therapy, have found to be quick response being coupled with synthetic medicines and surgical processes. But gradually, it was found to be associated with many side effects and costlier for common people. On the other hand Ayurvedic therapy is mainly based on the easily accessible indigenous plants which are free from side effects. Of course, the process of curing diseases is comparatively slow as it treats the body as a whole through natural means. Hence, in this modern age the Ayurvedic therapy has been adopted by the people belonging to different walks of life and even for beautification process by the aristocrat society. Therefore, there is necessity for proper identification of medicinal plants. So that this will provide better efficacy and side by side scientific names of the medicinal plants should be used in pursuance with provisions of International Code of Botanical Nomenclature for wider acceptance.

Materials and Methods

From the survey of various treatise like Bhava Prakash Nighantu (Chunekar, K.C 1982) Nighantu Adarsha (Vaidya, B.G., 1968, and 1985) some controversial drugs of Indian Medicine (Vaidya, B.G. 1982) Pharmacognosy of Indigenous drugs (Raghunathan and Mitra,1982), Pharmacognosy of Ayurvedic drugs (Iyer and Kolammal, 1951–1979), Data base on medicinal plants used in Ayurveda (Sharma, P.C. *et al*, 2000–2004) it has been observed that some medicinal plants have poly–botanical equivalents. Different regions of this country use different species under same name. For example Brahmi a potent memory booster plant has been used as *Bacopa monnieri* and *Centella asiatica*. Similarly *Eclipta prostrata, Wedelia chinensis* and *Limnophila rugosa* plants have been used for Bhringaraj. Shankhapuspi, Jeevanti, Pashanabheda, and Shalaparni are associated with medicinal plants such as *Canscora decussata, Evolvulus alsinoides, Convolvulus pluricaulis, Clitoria ternatea; Leptadenia reticulata, Holostemma annulare; Rotula aquatica, Aerva lanata, Kalanchoe pinnata, Polycarpea aurea* and *Desmodium gangeticum, D. polycarpum, Flemingia stricta* respectively. However, these medicinal plants have been collected by the traditional plant collectors basing on their sanskrit names or local names. Sanskrit names as cited in the ancient treatise are associated with brief descriptions while local names vary from one place to another and from one dialect to another. Consequently, the efficacy of Ayurvedic medicines prepared from such misidentified medicinal plants varies and decreases.

At present, medicinal plants have been scientifically identified by the help of Botanical keys and characters. For example, *Bacopa monnieri* and *Centella asiatica* (known as Brahmi) belong to family *Scrophulariaceae* and *Apiaceae* respectively. Hence, the floral structure of *B. monnieri* is gamopetalous, zygomorphic, ovary hypogynous, multiovulate while *C. asiatica* is polypetalous, actinomorphic, ovary inferior with few ovules. As regards Bhringaraj *Eclipta prostrata* and *Wedelia chinensis* belong to Asteraceae family. Therefore, in both the cases ray florets and disc florets are zygomorphic and discflorets are actinomorphic, ovary inferior, I-ovule with basal placentation. But another species *Limnophila rugosa* belongs to family scrophulariaceae, where there is no differentiation of floral structure. All are zygomorphic, ovary hypogynous, multiovulate. Under Shankhapuspi *Canscora decussata* belongs to Gentianaceae family. The floral structure is gamopetalous, zygomorphic, placentation parietal. But *Evolvulous alsinoides* and *Convolvulous pluricaulis* are of family convolvulaceae, hence gamopetalous, actinomorphic, ovary superior, few ovules with axile placentation; while *Clitoria ternatea* belongs to family Fabaceae and the floral structure is polypetalous, zygomorphic, ovary superior, marginal placentation. Similarly *Leptadenia reticulata* and *Holostemma annulare* known as Jeevanti both belong to family Asclepiadaceae. *L. reticulata* anthers without membranous tip, pollinia sessile, corolla velvate, greenish white and *H. annulare* anthers without membranous tip, pollinia often pedicelled, corolla pale-pink. In the name of Shalaparni three species namely *Desmodium gangeticum, D. polycarpum* and *Flemingia stricta* are used, all belonging to family Fabaceae. *D. gangeticum* leaves I-foliolate, joints about as broad as long while in *D. polycarpum* leaves 3-foliolate, joints scarcely longer than broad, In *Flemingia stricta* plants shrubby, leaves 3-foliolate.

Besides scientific identification of the medicinal plants as available in Indian Medicinal Plants (Varier P.S. 1995-1996), Pharmacognosy of Indigenous drugs (Raghunathan and Mitra, 1982), Pharmacognosy of Ayurvedic Drugs (Iyer and Kolammal, 1951-1979) and Medicinal Plants of India (Satyavati G.V. *et al.*, 1976-1987) the nomenclature of medicinal plants is also equally important for International communication as per International Code of Botanical Nomenclature (Tokyo Code, 1994).

Sometimes misidentification of the plants also posses some problem. For example *Saraca asoca,* a potential medicinal plant for gynaecological disorders has been identified as *Saraca indica* which is not available in the Indian subcontinent. It has been proved that all the plants named as 'Ashoka' in India is *Saraca asoca*.This should be strictly followed while naming the constituents of Ayurvedic medicines prepared out of Ashoka.

Discussion

Due to unscientific and indiscriminate collection of medicinal plants many plants are becoming rare and endangered. Such plants include *Rauvolfia serpentina, Plumbago indica, Gloriosa superba, Saraca asoca, Chlorophytum arundinaceum, Symplocos racemosa* and *Pterocarpus santalinus* etc. Reserpine is the main alkaloid present in the mature root of *Rauvolfia serpentina.* This plant is uprooted by the plant collectors for supplying mature roots to the drug manufacturing industries, while the seeds of *R. serpentina* is less viable and therefore, the natural regeneration of this species is gradually decreasing and it has become endangered. Similarly, *Gloriosa superba* rhizome (colchine) is used for medicine and whole plant is uprooted causing its rarity. The bark of *Saraca asoca, Symplocos racemosa*; roots of *Plumbago indica, Chlorophytum arundinaceum* and the heart wood of *Pterocarpus santalinus* are used in Ayurvedic therapy. Destructive harvesting process resulted the shrinkage of medicinal plant population.

Under such circumstances adequate strategies should be adopted through latest modern technology like micropropagation method. Due to rarity of *R. serpentina* root quiet often the roots of *R. tetraphyla* (*R. canescens*) is used as substitute since the plant is easily available because its production of more seeds and viability. However, the alkaloidal content is comparatively less in roots of *R. tetraphyla.* Similarly due to non availability of *Saraca asoca* bark *Polyalthia longifolia* bark is adulterated which is locally known as Ashok.

Taking into account of all these aspects Ayurvedic therapy may be modernized so that its efficacy and popularity will be gained.

Acknowledgement

The first author (BNM) is thankful to Dr. J. Nath, Principal Govt. Ayurved College, Bolangir for his constant encouragement.

References

Borthakuar, S K (2004) Problems and prospects of Medicinal plants as an organised sector in Assam. In: National Seminar on Medicinal Plant in NE Region. P-65-67, held on Oct.07 to 10 at Guwahati, Assam.

Brahman, M (2004) Herbal drugs: Origin and importance in phytotherapy. e-planet journal 2(1): 64-69.

Chunekar, K C (1982) Bhava Prakash Nighantu, Chaukhambha Bharati Academi, Varanasi.

Chopra, R N, S L Nayar and I C Chopra (1956) Glossary of Indian medicinal plants, New Delhi.

Garcia d, Orta (1563) Os Colloquius, Madrid.

Greuter, W *et al.* (1994) International Code of Botanical Nomenclature (Tokyo Code) Koeltz Scientific Books. D-61453, Kanigstein, Germany.

Hooker, J D (1872-1897) The Flora of British India Vol. I-VII, London.

Iyer, K N and M Kolammal (1951-1979) Pharmacognosy of Ayurvedic drugs series 1-10, University of Kerala, Trivandrum.

Kirtikar, K R and B D Basu (1918) Indian Medicinal Plants Vol. I-IV, Allahabad.

Nadkarni, A K (1976) Indian Materia Medica Vol.-I, Popular Prakashan, Bombay.

Naik, V N (2004) Taxonomy and Conservation of Indian Medicinal Plants in: Biodiversity and Sustainable utilization of Biological resources (ed T.R. Sahu and P.K. Sahu). Scientific Publishers (India) Jodhpur.

Pullaiah, T (2002) Medicinal Plants in Andhra Pradhesh (India). Regency Publication. New Delhi.

Raghunathan, K and Miss Roma Mitra (1982) Pharmacognosy of Indian Drugs, Vol. I and II, Central Council for Research in Ayurveda and Siddha, New-Delhi.

Rheede, H A Van (1678-1703) Hortus Indicus, Malabaricus, Vol. I-XII, Amsterdam.

Roxburgh, W (1814) Hortus Bengalensis, Calcutta.

Roxburgh, W (1832) Flora Indica, Vol. I-II, Calcutta.

Satyavati, G V, Raina, M K and M Sharma (1976-1987) Medicinal Plants of India Vol. I and II, New-Delhi.

Satoskar, R S and S D Bhandarkar (1980) Pharmacology and Pharmacotherapeutics Part I and II. Popular Prakshan, Bombay.

Sharma, P C, M B Yelne and T J Dennis (2000-2004) Database on Medicinal Plants Vol. I-IV. Central Council for Research in Ayurveda and Siddha, New-Delhi.

Tripathy, K D (2003) Essentials of Medical Pharmacology. Jaypee Brothers Medical Publishers (P) Ltd. New Delhi.

Vaidya, B G (1968) Nighantu Adarsha (Poorvardha-Hindi) Chaukhambha Vidyabhavan, Varanasi.

Vaidya, B G (1982) Some controversial drugs of Indian medicine, Chaukhambha Orientalia, Varanasi.

Vaidya, B G (1985) Nighantu Adarsha (Uttarardha-Hindi) Chaukhambha Bharati Academy, Varanasi.

Varier, P S (1995-1996) Indian medicinal plants Vol. I-V. Arya Vaidya Shala, Kottakal, Orient Longman.

Watt, G (1887-1893) Dictionary of economic products of India, Bombay.

Chapter 15

Ethnomedicinal Plants of Hemgir Forest Range of Sundargarh District in Orissa, India

B.L. Bohidar [1] and L.M. Behera [2*]

[1]*Department of Botany, Womens' College, Jharsuguda – 768 201, India*

[2]*P.G. Department of Botany, G.M. (Auto) College, Sambalpur – 768 004, India*

ABSTRACT

An ethnobotanical study was carried out in the major tribal belts of Hemgir, Kanika, Daghora, Rohini and Kodbahal forest sections of Hemgir forest range in Sundergarh district of Orissa. The present study enumerates the uses of 25 plants (mostly trees) belonging to 16 families by the ethnic groups, which include the *Gond, Munda, Pobia, Kondh, Khadia, Bhuiyan, Kol, Kisan and Oram* tribes inhabiting the area.

Keywords: *Ethnomedicine, Hemgir, Sundergarh, Tribals.*

Introduction

Sundergarh, one of the North-Western districts of Orissa lies between 21° 35′ N and 85° 42′ N latitude and 83° 32′ E and 85° 22′ E

* Email: lambehera@sancharnet.in; Ph: (0663) 2400106, (M) 9437050526.

longitudes covering an area of 9712 sq kms. The district is bound on the North by Ranchi district of Jharkhand; on the East by Singhbhum district of Jharkhand and Keonjhar districts of Orissa; on the South by Deogarh, Sambalpur and Jharsuguda districts of Orissa and on the West by Raigarh district of Chhatisgarh. Geographically, the district is an undulating table land of different elevations broken-up by rugged hill ranges and cut-off by torrential hill streams and the river Ib and Brahmani.

Hemgir, one of the five forest ranges of Sundergarh forest division is situated in South-Western part of Sundergarh and lies between 83° 32′ N and 83° 57′ N latitude and 21° 45′ E and 22° 2′ E longitudes covering an area of 26679.16 hectacres chiefly constituting the sub-tropical dry deciduous mixed "Sal" forest dominated by *Shorea robusta*. According to the 2001 census Hemgir range has highest concentration (63.7 per cent) of tribal population in the district mostly residing in villages. The *Gond, Munda, Pobia, Oram, Khadia, Bhuiyan, Kol, Kisan and Kondh* are the dominant tribals of this range along with non-tribal people. There are also the traditional medicine practitioners commonly known as 'Vaidya' and 'Kabiraj' who utilize a number of medicinal plants for cure of various diseases. A number of reports are available on ethnobotany of different parts of Orissa (Jain 1970, 71 and 1973; Rai Choudhary *et al.*, 1975; Saxena and Brahmam, 1975; Subudhi and Choudhury 1985; Girach *et al.*, 1985; Mishra 1992; Sahoo and Mudgal, 1995). However elaborate ethnomedicinal survey has not yet been done, so far from Sundergarh district. With this background the present study was undertaken in Hemgir forest range of the district which is dominated by tribal populations who depend on the plants for their primary health care.

Methodology

The authors have conducted an extensive field survey in the tribal belts and selected villages of the study area for collection and record of plant species of ethnomedicinal value during September 1998 to February 2000. The tribals possess fairly good amount of knowledge about the diagnosis and treatment of several diseases usually encountered by them in their day to day life. Information on curative values of different plant species has been gathered through personal interviews with tribal men, women and traditional medicine practitioners of the region. The collected plant specimens were identified with the help of floras (Haines, 1978; Saxena and

Brahmam, 1994) and have been preserved in the herbarium of Botany Department, Womens' College, Jharsuguda, Orissa as voucher specimens. To ascertain the uses of these medicinal plants, the earlier published scientific literature sources (Jain 1991; Kirtikar and Basu 1991; Chopra *et al.*, 1992; Nadkarni 1998) were referred. Information gathered during the present investigation are enumerated with their botanical names in alphabetical order followed by family names in brackets and local names in parentheses and localities of collection with collection number in bracket followed by medicinal uses of each species.

Observation

Information on 25 medicinal plants commonly used by the tribals and traditional medicine practitioners of Hemgir forest range are described as follows:-

Aegle marmelos (L) Correa. (Rutaceae), 'Bel', Arjunda (12)

Unripe fruits are cut into pieces and sun dried (*Bel Sutha*). It is rubbed on a stone and mixed with two spoonful of honey. It is administered orally twice a day for 3 days to cure blood dysentery of children.

Seven leaves of 'Bel' (*Aegle marmelos*) and seven seeds of black pepper (*Piper nigrum*) are to be chewed like betel (*Piper betel*) with rock salt for 21 days every morning in empty stomach to cure gastro-intestinal disorders.

Ailanthus excelsa Roxb. (Simaroubaceae) 'Mahanimb', Bendrichuan (16)

Five to seven leaflets are boiled in 300 ml of water till it becomes 100ml. It is strained and the decoction is taken orally (10 ml per dose, twice a day) for 3 days to cure malaria.

Azadirachta indica A. Juss (Meliaceae) 'Nimb', Kalmegha (24)

Bark is pounded with equal amount of turmeric (*Curcuma longa*) and the paste is applied all over the body to prevent from scabies and ring worm.

Sundried vegetative bud powder is taken daily for 10-15 days especially during February to April to prevent measles, chickenpox and smallpox.

Bombex malabarica D. (Bombacaceae) 'Semel', Kalmegha (35)

Gum of the plant with dried mango (*Mangifera indica*) seeds are taken in equal quantities and powered. The powder is mixed with honey and made into tablets. Two tablets twice a day for 4 days cures painful menstruation.

Thorn is rubbed on a stone with cow milk and the paste obtained applied externally on face daily for one month continuously cures pimples as well as removes the mark of pimples.

Buchanania lanzan spreng. (Anacardiaceae) 'Char', Beldihi (41)

Stem bark and root are made into a paste. The paste is applied to cure chronic wounds

Butea monosperma Lamk, (Fabaceae) 'Phalsa', Kutrajore (31)

Gum is made into powder and mixed with country liquor. It can be given just for 3 days after menstruation to prevent pregnancy.

Root is ground with rice washed water and the paste is applied on neck to cure goitre.

Careya arborea Roxb. Lecythidaceae 'Kumbhi', Majhapara (52)

The bark extract with paste of the fruit 'Jaiphal' (*Myristica fragrant*) is given in empty stomach once daily for 3 days to cure blood dysentery.

Fruit juice with 'Mishri' (Sugar candy) water given 3 times a day for 7 days to cure dysuria.

Cassia angustifolia Vahl. (Caesalpiniaceae) 'Sunamukhi', Budelkani (51)

Dried leaves of Sunamukhi (*Cassia angustifolia*), dried fruits and seeds of 'Vidanga' (*Emblica ribes*) are powdered. The powered mixture is given with lukewarm water for chronic constipation and worm infestation in children.

***Cassia fistula* L. (Caesapiniaceae) 'Sunari', Kutrajore (43)**

Juice of tender leaves are applied on the skin affected by eczema. The paste of the bark is taken orally by women for abortion. The juice of the seed is taken to cure amoebic dysentery.

***Diospyros melanoxylon* Roxb. (Ebeneaceae) 'Kendu', Lipospali (15)**

Paste of the bark with rice washed water is given two times a day for 3 days against diarrhoea.

***Emblica officinalis* L. (Euphoribiaceae) 'Aonla', Tilia (61)**

Dried fruits of 'Aonla' (*Emblica officinalis*) with 'Beheda' (*Terminalia belerica*) and 'Harda' (*Terminalia chebula*) are soaked overnight in a glass of water and taken early in the morning to cure constipation.

Fresh leaf juice is mixed with filtered limewater. The mixture is given twice a day for 3 days to cure dysentery of children.

***Ficus benghalensis* L. (Moraceae) "Bar" Julumbahal (73)**

Paste of the apical part of prop root with rice washed water is given two times a day for 7 days to cure a type of dysentery called '*Patar Jhada*' of children.

***Madhuca India* Roxb. (Sapotaceae) 'Mahul' Badrampia (63)**

Edible oil obtained from seeds is used in skin diseases. Honey obtained from flower is applied in the eyes to cure redness.

***Gardenia gummifera* L. (Rubiaceae) 'Kurlu' Kusmel (48)**

The bark of 'Kurlu' (*Gardenia gummifera*), fruit rind of 'Beheda' (*Terminalia belerica*) and 'Haldi' (*Curcuma longa*) are powdered together. The powdered mixture is taken twice daily with honey for intestinal worms.

The resin extract from bark is mixed with warm mustard (*Brassica campestris*) oil and 'Aphu'(*Opium*). This mixture is filtered to get a clear oil. This oil is used for massage for arthritis and backache.

***Mangifera indica* L. (Anacardiaceae) 'Aam' Tumapali (09)**

The juice of fresh bark is mixed with egg mucilage and a little opium (*Papaver somniferum*). The mixture is given two times daily to cure bleeding piles.

***Pongamia glabra* Vent. (Fabaceae) 'Karanjo', Khursi (72)**

The oil extract from seed is warmed and applied locally in case of boils and wounds.

***Pterocarpus marsupium* Roxb. (Fabaceae) 'Bija', Dheknapani (74)**

The bark extract is taken with sugarcane (*Saccharum officinarum*) juice to cure 'red-urine' diseases. The juice of the bark is applied on the tongue in tongue diseases.

***Schleichera oleosa* willd. (Sapindaceae) 'Kusum', Dungridihi (08)**

Oil extract from seed is mixed with garlic (*Allium sativum*) and leaf juice of 'Datura (*Datura alba*) and are cooked together, and the oil thus obtained is used as a massage oil for chronic rheumatism. Oil extract from seed is applied to cure eczema and boils.

***Shorea robusta* Gaertn. f (Dipterocarpaceae) "Sargi" Tumapali (57)**

Latex is cooked in mustard oil and turmeric powder. The ointment obtained is applied to cure infections of skin between the toes locally know, a 'Keintra'.

***Streblus asper* Lour. (Moraceae) 'Sahada', Surulata (82)**

The stem bark of 'Sahada' (*Streblus asper*) root bark of 'Kure' (*Holarrhena antidysenterica*) and 'Dhatuk' (*Woodfordia fruticosa*) are powdered together. The powdered mixture is taken twice a day with honey for one month to cure leucorrhea.

***Syzygium cumini* L. (Myrtaceae) 'Jamu' Julumbahal (07)**

Seeds of 'Jamu' (*Syzygium cumini*), 'Juani' (*Carum copticum*), and 'Nim' (*Azadirachta indica*) and whole plant of 'Bari Anola' (*Phyllanthus asperulatus*) are powdered. The mixture is taken regularly with juice of 'Karla' (*Momordica charantia*) and tender leaves of 'Nim' (*Azadirachta indica*) in empty stomach for diabetes.

Tamarindus indica L. (Caesalpiniaceae) 'Tetel', badrampia (88)

Extract of old *Tamarindus indica* fruits mixed with old jaggery is given to ladies for curing *menorrhogia* and its related complaints like weakness, backache and anaemia.

Terminalia arjuna Roxb. 'Kaha', (Combretaceae) Dungridihi (77)

The fresh bark juice of 'Kaha' (*Terminalia arjuna*) with bark juice of blackberry (*Sygygium cumini*) and juice of zinger (*Zingiber officinalis*) given every morning in empty stomach to cure diabetes.

Bark ash mixed with warm 'Nim' (*Azadirachta indica*) oil is applied on the skin affected by eczema.

Terminalia chebula Retz. (Combretaceae) 'Harda', Julumbahal (84)

The dried fruits of 'Harda' (*Terminalia Chebula*), 'Beheda' (*Terminalia bellirica*) and 'Aonla' (*Emblica officinalis*) are powdered together. The powder is taken with lukewarm water for the treatment of indigestion, acidity and constipation. The dried fruit is roasted and chewed as such for one month for dry cough.

Terminalia bellirica Roxb. (Combretaceae) 'Beheda', Arjunda (83)

The fruit pulp mixed with pepper (*Piper nigrum*), 'Bacha' (*Acorus calamus*) and honey is used to cure sore throat, hoarseness of voice and cough.

Discussion

The tribals *Gond, Munda, Pobia, Kondh, Bhuyan, Kol, Kisan and Oram* live in Sundergarh district. They have good knowledge of medicinal uses of the plant species available in the region. But they are conservative in nature and do not want to share their age-old traditional knowledge with other people. During the course of the study it was observed that many of the experienced traditional medicine practitioners who are already dead had never disclosed their knowledge of medicinal uses of plants to others. However, after developing intimacy with some of the experienced medicine men and other traditional healers who are alive, some information on medicinal uses has been collected and is presented in this chapter. Traditional use of medicinal plants is slowly disappearing due to destruction of forests for urbanization and industrialization in this district. Hence it is time to take appropriate measures for *in situ* and *ex situ* conservation of the useful plant species. The chapter presents a brief account of the uses of some of the ethonomedicinal plants used against various diseases by the people of Hemgir forest range and urges the need for further investigation on biochemical and pharmaceutical aspects of these ethnomedicinal plants.

References

Bhattacharjee SK (2000) Hand Book of Medicinal Plants. Pointer Publisher Jaipur, 474 P.

Chopra RN Verma BS and Chopra IR (1992) Supplement to the Glossary of Indian Medicinal Plants, CSIR, New Delhi, 119 P.

Das PK and Mishra MK (1988) Some ethnobotanical plants of Koraput district, Orissa. Ancient Science of Life. 8:60-67.

Hains HH (1978) The Botany of Bihar and Orissa, Bisen Singh and Mahendrapal Singh, Dehradun, 1350 P.

Jain SK (2001) Dictionary of Indian Folk-Medicine and Ethnobotany. Deep Publications, New Delhi, 311 P.

Jain SK (2001) Indian Medicinal Plants, National Book Trust, India, New Delhi 216 P.

Kirtikar KR and Basu RB (1999). Indian Medicinal Plants, International Book Distributors, Dehradun, 2761 P.

Mudgal V and Pal DC (1980). Medicinal plants used by tribals of Mayurbhanj, Orissa. Bull. Bot. Surv. India 22:59-62.

Nadkarni KM (1998) Indian plants and drugs with their medicinal properties and uses. Asiatic Publishing House, New Delhi, 450 P.

Satpathy KB and Panda, PC (1992). Medicinal uses of some plants among the tribals of Sundergarh district, Orissa. J Econ. Tax. Bot. Addl. Ser. 10:241-249.

Saxena HO and Brahman M (1994). The flora of Orissa. Orissa Forest Development Corporation., Bhubaneswar, 2918 P.

Chapter 16

Nutrient Status of Water Bodies at Burla, Orissa

Khageshwar Pradhan*

Research Scholar,

Department of Environmental Sciences, Sambalpur University,

Jyoti Vihar, Orissa – 768 019

ABSTRACT

Over last few decades the quality of freshwater is deteriorating due to increased growth in population, urbanization and industrialization. Small water bodies like pond and ditch are considered to be more important in rural areas which serve for even domestic consumption besides its normal use for bathing, washing etc. as well as waste water disposal site. These activities provide a continuous supply of nutrients leading to deterioration in quality of water. Burla is a small town in the district of Sambalpur having an area of 19.42 sq.km. The water supply by the Government is limited to the employees only. However, around 30 per cent of the population depends upon their own water source including the pond water during lean season. Therefore, it was necessary to generate information on the quality of four major ponds of the locality and to recommend for its appropriate use. The ponds selected are (*i*) Sadeipali pond (*ii*) University pond (*iii*) Katapali pond and (*iv*) Golgunda pond. Fourteen parameters were selected

* *Ph: (M) 09437218993; E-mail: kp_envsc@rediffmail.com.

for analysis of water quality including the nutrients like phosphate and nitrate. The ponds were ranked in order of pollution and it was found that out of the four ponds University pond is highly polluted followed by Golgunda, Sadeipali and Katapali pond. This is largely due to influx of nutrients (phosphate, nitrate, sodium and potassium) in run off from catchment areas leading to the process of eutrophication. Therefore it is desirable to construct bunds surrounding the pond to restrict the influx of nutrients and prevent the use of pond for washing and bathing of domestic cattle.

Keywords: *Deterioration, Nutrient, Pond ecosystem.*

Introduction

Water is the elixir of life without which no biota could survive in biosphere. An indisputable and inseparable bond exists between fresh water bodies and human beings (Kumar, 2002). It is an essential constituent of all animals and vegetable matter and forms about 75 per cent of the matter of earth crust. The Indian society is gradually changing from an agriculture-based society to agriculture industry-based society. Thus increasing amount of fertilizers, industrial wastes, sewages, detergents, insecticides etc. are being washed in to water, changing the natural character of these water bodies. Over last few decades due to increased human population and man-made condition, the water quality is deteriorating everywhere. It is therefore necessary that the quality of drinking water should be maintained and checked at regular time interval (Prajapati and Mathur, 2002). The quality of water has generally been signified by the origin and the extent to which it is contaminated on the way by anthropogenic means (Haque *et al.*, 2002). And also water serves as a medium in which the plants and animals *of* different water bodies are greatly affected due to discharge of domestic, municipal, industrial and other several factors (Panda *et al.*, 1991; Sahu *et al.*, 1994 and 1995 and Batcha, 1998) during day and night which may be fatal for aquatic life.

In a water body, the nutrient content initiates and reflects the growth rate of the plants (hydrophytes). The gradual load of nutrients in an aquatic body tends it to the eutrophication stage (Pinckney *et al.*, 2001). Eutrophication is a process that can be defined as an increase in the rate of supply of organic matter (OM) and nutrients to an ecosystem. The main chemical constituent of OM is carbon (C), and therefore rates of eutrophication are expressed in units of C per area per unit time. OM occurs in both particulate and dissolved forms. Autochthonous OM originates outside the water body, whereas autochthonous OM is generated within the system, mostly by primary producers or by benthic regeneration of OM. The supply rates of limiting nutrients regulate phytoplankton productivity that contributes to inputs of autochthonous OM. The trophic status of water body is often based on eutrophication rates and can be categorized as oligotrophic (<100 g C m^{-2} y^{-1}), mesotrophic (100–300 g C m^{-2} y^{-1}), eutrophic (300–500g C m^{-2} y^{-1}) or hypertrophic (>500 g C m^{-2} y^{-1}). Ecosystem responses to eutrophication depend on both export rates (flushing, microbially mediated losses through respiration, and denitrification) and recycling/regeneration rates within the aquatic system. The mitigation of the effects of eutrophication involves the regulation of inorganic nutrient (primarily N and P) inputs into receiving waters. Appropriately scaled and parameterized nutrient and hydrologic controls are the only realistic options for controlling phytoplankton blooms, algal toxicity, and other symptoms of eutrophication in estuarine ecosystems.

Pond Ecosystem

Ponds are considered to be one of the most important water resources for man particularly in rural areas. These are the main sources of freshwater, meeting the multifarious demands of human population and are exploited for almost every use. In the early days, when the population was sparse, waters were able to cleanse themselves by way of their inherent self-purification capacity (Goel, 1997). Natural process like decomposition of organic wastes, may also contribute to pollution. However, these processes takes place at a very slow speed and provide enough time for the waters to cleanse themselves. On the contrary, through various activities we are adding wastes in such huge quantities that the natural cleansing process is unable to cope up with the load. Modern civilization with its rapid increase in population and industrialization has led to an accelerated degradation of the freshwater systems. Urbanization aggravates this situation by the direct discharge of sewage and industrial wastes. The water bodies, thus receive contaminants ranging from silt to pesticides, and from industrial effluents to sewage materials

(Mukherjee, 1997). Such unwise exploitation disturbs the natural harmony of aquatic ecosystem (Singh *et al.*, 1982; Sinha and Kumar, 1983). There are a lot of evidences on the progressive deterioration of the water quality of various water bodies due to disposal of wastes in them (Saxena and Chauhan, 1993; Kandhasamy and Santhaguru, 1994; Pervez and Pandey, 1994; Sahu *et al.*, 1995; Jha *et al.*, 1997; Nair and Ganapati, 1997).

Waste Water Generation in India

A survey carried out by Central Pollution Control Board (CPCB 1989) on the status of generation, collection, treatment and disposal of waste water in 212 class-1 cities covering 65 per cent of Indian population indicates a total generation of 12,145 million litres of waste water per day (MLD) out of which only 2633 MLD (22 per cent) are collected through sewage system and rest are discharged to land or water without treatment. From the total collection of 2633 MLD through sewage system 895 MLD gets primary and secondary treatment. In the absence of tertiary treatment the released waste water is capable of causing eutrophication in water bodies. In India some 5000 large and medium industrial units are responsible for pollution of our water bodies. But their contribution is less than 10 per cent pollution of fresh water in India is largely due to municipalities. For example, the industries of Bombay account for only 13 per cent of total waste dumped in to water bodies and in Calcutta are 11 per cent. In Delhi river Yamuna takes about 200 million litres of untreated human waste every day. In Varanasi, 225 MLD sewage is generated which is mixed with industrial effluent containing toxic metal and hazardous chemical, pesticides, insecticides etc.

Composition of Domestic Sewage Water

The pollution of water bodies in developing countries is mainly due to the discharge of domestic sewage containing various dissolved substances. The untreated wastes deteriorate the water quality and affect adversely the aquatic flora. Composition and qualities of wastewater generated from house hold activities differ from region to region and from season to season. Some of the raw sewage is rich source of nutrients to plants and fishes. A host of them could cause toxicity and become detrimental when present in amount larger than tolerable limit.

Objectives of the Present Investigation

Since there is growing demand among different sectors of the community for the use of water for drinking, irrigation, industries etc. water quality is becoming important in the present public health set up for our community. With enormous growth of population and the advent of industrialization, ecosystems are getting adversely influenced by the activities of man and his livestock. The community ponds of Burla are very extensively used for bathing, utensil and cloth washing, cattle bathing and the banks are used as open air latrines by people. The ponds now-a-days are used for cleaning transport vehicles. Hence it is hypothesized that the water may not be safe for drinking purpose and possibly for any form of human use. Therefore it was believed that this work would provide quantitative information on the pond systems so that remedial measures can be taken and ponds can be used for other productive purposes. Keeping these aspects in mind, Burla town was visually surveyed and it was observed that four ponds are extensively subjected to human use.

Hence considering accessibility and extent of use by people, these ponds were chosen for detailed physico-chemical study for three seasons in a year.

Study Site

Burla is situated at a distance of 15 kms north-west from Sambalpur. It is a small town of Sambalpur district having an area of 19.42 sq.km. This town has a medical college and hospital, Engineering College and a University. It lies between 21° and 22°N latitude and 83° and 84°E latitude and 160 mts above the sea level. This town has a population of 39188, 2001 census. This town is bounded by the Hirakud Dam reservoir, except the western side. Very recently the Notified Area Council (NAC) has made provisions for drinking water. But the town water supply is very insufficient hence the waters of the ponds and wells are utilized by the people for various purposes. This is a double crop agricultural based town. Paddy (rice) and vegetables are the main agriculture products of this area. As the canal facility is available for irrigations farmers are interested for double crop. In the sense it needs more fertilizer and pesticides. Thus the run-off water from both agriculture and non-agriculture field enter into the water bodies.

In 2001 census, the population of Burla is 39188. Although most of the inhabitants of Burla are provided with tap water, supply water and tube well facilities for domestic consumption, the ponds were developed in and around the Burla for domestic consumption as well as for washing, bathing etc. So several peoples are dependent upon these ponds. Like in the Sadeipali pond, University pond, Katapali pond and Golgunda pond, the no of dependent people are 3500, 2800, 2500 and 3000 respectively. Therefore it is necessary to generate information on the quality of the ponds water and to recommend for its appropriate use.

Materials and Methods

Water samples were collected in pre-monsoon, monsoon and post-monsoon season from ponds at surface and bottom of the water level and kept in previously washed and dried plastic containers. The containers were rinsed thoroughly with sample water, and then the collected samples were taken. The samples were collected through the water sampler from the desired depth. Some important physico-chemical parameters were analyzed. Temperatures of all the samples were recorded immediately after collection. D.O. was fixed at the sampling site and determined in the laboratory by using Winkler's method. Water Electrical Conductivity, Total Dissolved Solid and Sodium and Potassium were estimated by using Systronics pH meter, Conductivity meter, TDS meter and Flame Photo meter respectively. The Chemical Oxygen Demand was estimated by Potassium dichromate method. The nitrate and phosphate were done by phenol disulphonic acid and stannous chloride method respectively. The remaining all the parameters were conducted by manual titrimetric methods.

Results

The seasonal variations in physico-chemical characteristics of the four ponds (Sadeipali pond, University pond, Katapali pond and Golgunda pond) during pre-monsoon, monsoon and post monsoon have been summarized in respective tables. The interpretation of data has been made with the help of statistical tools. The variation in seasonal average in temperature variation at the four ponds both at surface and bottom were found (Table 16.1). In general the temperature of both surface and bottom water was less in post-monsoon and more in pre-monsoon period. The average temperatures of four ponds were 31.05, 30.01, 29.7 and 28.6°C in Sadeipali pond, University pond, Katapali pond and Golgunda pond respectively. The variations in temperature between different seasons were highly significant (Table 16.3). The water of Sadeipali pond was relatively more alkaline than the others and in general the surface was more alkaline than the bottom water. Thus the surface and bottom water of all the ponds were alkaline in nature and in all the ponds pH content in both surface and bottom were maximum in pre-monsoon (Table 16.2). Anova (Table 16.3) reveals a significant difference in pH of surface water between different seasons. But in case of bottom water the difference is significant between the ponds but not the seasons. The bottom water was marginally low in conductivity than the surface water. In all the ponds Electrical Conductivity and TDS (Total Dissolved Solids) in both surface and bottom water was maximum in monsoon and minimum in pre-monsoon. The analysis of variance (Table 16.3) shows a significant difference in Electrical Conductivity in the surface water between the ponds as well as seasons. Where as in the bottom water, the difference is significant between the ponds, but not between seasons. The difference in TDS contents between the seasons for both surface and bottom water where as the difference was not significant between the ponds.

The dissolved oxygen (DO) content of the surface water ranged from a minimum of 2.08±0.245 mg/l^{-1} (pre-monsoon) in University pond to a maximum of 8.27±0.199 mg/l^{-1} (pre-monsoon) in Golgunda pond (Table 16.1). Hence the water of University pond was relatively more polluted than the others. The bottom water was low in DO content than the surface water. In all the ponds, DO content in both surface and bottom water was maximum in post-monsoon and minimum in pre-monsoon. The analysis of variance (Table 16.3) showed a significant difference in DO content between the ponds but not the seasons in the surface water where as in the bottom water the difference was significant both between the ponds as well as seasons. The higher and lower Chemical Oxygen Demand (COD) values were found in pre-monsoon and post-monsoon respectively in both surface and bottom water of all the ponds. However, analysis of variance (Table 16.3) reveals no significant difference in COD between the ponds as well as the seasons for both surface and bottom water.

Table 16.1: Summary of Mean Temperature (єC), pH, Electrical Conductivity (EC mmho/cm), Total Dissolved Solid (TDS mgl^{-1}), Dissolved Oxygen (DO mgl^{-1}), Chemical Oxygen Demand (COD mgl^{-1}), Phosphate (PO_4 mgl^{-1}), Nitrate (NO_3 mgl^{-1}), Cl (Cl mgl^{-1}), Free carbon Dioxide (CO_2 mgl^{-1}), Alkalinity (mgl^{-1}), Hardness (mgl^{-1}), Sodium (Na mgl^{-1}) and Potassium (K mgl^{-1}) were Recorded within the Four Selected Ponds with Surface and Bottom of the Water Level in Three Seasons (Pre-monsoon, Monsoon and Post-monsoon).

		Physico-chemical Parameters													
		Temp.		*pH*		*EC*		*TDS*		*DO*		*COD*		*PO_4*	
		Mean	*SD*	*Mean*	*SD*	*Mean*	*SD*	*Mean*	*SD*	*Mean*	*SD*	*Mean*	*SD*	*Mean*	*SD*
							Sadeipali Pond								
Surface	A	38.8	0.081	8.13	0.008	0.33	0.001	0.13	0.001	3.68	0.13	163.99	47.367	630	40
	B	25.5	0.094	8.09	0.004	0.64	0.002	0.64	0.002	6.19	0.15	82.32	2.160	430	20
	C	29.9	0.047	8.04	0.032	0.49	0.001	0.43	0.001	7.47	0.075	43.97	1.886	90	40
Bottom	A	37.5	0.205	7.95	0.008	0.32	0.001	0.16	0.001	2.38	0.245	109.32	5.657	760	0
	B	25.2	0.081	794	0.004	0.62	0.001	0.67	0.001	5.01	0.075	62.65	19.686	630	40
	C	29.7	0	7.96	0.016	0.48	0.002	0.44	0	6.77	0.075	28.07	1.819	220	40
							University Pond								
Surface	A	37.4	0.047	7.76	0.004	0.93	0.002	0.30	0.003	2.08	0.226	166.65	4.989	1540	60
	B	24.7	0.081	7.73	0.016	2.02	0.002	0.51	.0.001	3.2	0.13	125.32	5.657	1235	35
	C	28.7	0.141	7.69	0.014	1.73	0.001	0.49	0.002	4.96	0.569	61.31	1.886	925	25
Bottom	A	36.2	0.047	7.72	0.004	0.86	0.001	0.36	0.002	2.24	0.226	143.99	7.542	1735	65
	B	24.3	0.047	7.72	0.008	1.99	0.002	0.52	0.001	2.67	0.15	95.99	4.989	1380	0
	C	28.8	0.047	7.63	0.008	1.39	0.002	0.49	0.001	4.91	0.075	49.97	1.886	980	0

Contd...

Table 16.1–Contd...

		Physico-chemical Parameters													
		Temp.		*pH*		*EC*		*TDS*		*DO*		*COD*		*PO_4*	
		Mean	*SD*	*Mean*	*SD*	*Mean*	*SD*	*Mean*	*SD*	*Mean*	*SD*	*Mean*	*SD*	*Mean*	*SD*
								Katapali Pond							
Surface	A	35.5	0.047	8.07	0.004	0.54	0.001	0.23	0.002	7.84	0.130	89.32	0	1235	35
	B	25.2	0.047	7.97	0.008	0.83	0.001	0.47	1	6.93	0.075	53.32	5.657	965	15
	C	28.9	0.047	7.93	0.008	0.71	0.002	0.37	0.001	8.21	0.272	87.97	3.771	715	45
Bottom	A	35.3	0.047	8.09	0.004	0.46	0.001	0.20	0.009	5.87	0.150	67.99	1.886	1950	20
	B	24.9	0.047	7.96	0.03	0.83	0.003	0.49	0.001	6.29	0.150	39.99	1.886	1250	20
	C	28.4	0.047	7.90	0.004	0.68	0.002	0.38	0.002	7.52	0.130	81.31	1.886	900	50
								Golgunda Pond							
Surface	A	35.8	0.047	7.95	0.003	0.53	0.001	0.22	0.001	8.27	0.199	90.65	4.989	90	40
	B	24.8	0.047	7.92	0.008	1.21	0.001	0.52	0.001	4.69	0.199	98.65	3.771	255	125
	C	29.2	0.094	7.89	0.021	0.93	0.001	0.39	0.001	6.67	0.075	41.31	1.886	195	65
Bottom	A	*	*	*	*	*	*	*	*	*	*	*	*	*	*
	B	24.6	0.081	7.72	0.009	1.12	0.001	0.55	0.001	3.79	0.150	82.65	1.886	215	165
	C	28.6	0.047	7.9	0.016	0.64	0.002	0.41	0.001	6.24	0.130	23.97	3.771	100	0

Contd...

Table 16.1–Contd...

		Physico-chemical Parameters													
		NO_3		Cl		CO_2		Alkalinity		Hardness		Na		K	
		Mean	*SD*	*Mean*	*SD*	*Mean*	*SD*	*Mean*	*SD*	*Mean*	*SD*	*Mean*	*SD*	*Mean*	*SD*
							Sadeipali Pond								
Surface	A	4.21	0.01	31.94	4.512	2.03	0.047	73.33	1.885	407.33	0.943	65.43	0.262	185.73	0.124
	B	3.85	0.05	148.96	5.013	4.17	0.047	6.67	0.943	535.33	0.943	63.67	0.094	234.57	0.205
	C	2.41	0.01	112.53	4.679	4.67	0.094	58.67	0.943	545.33	1.886	63.53	0.094	209.77	0.59
Bottom	A	4.40	0	31.94	4.512	1.57	0.047	76	1.632	406.67	3.399	49.6	0.061	156.8	1.067
	B	4.10	0.1	127.69	5.013	3.47	0.094	5.33	0.943	541.33	1.943	49.97	0.262	261.6	1.355
	C	2.61	0.01	99.29	4.679	3.93	0.124	54	2.828	560	1.633	48	1.202	224.37	0.047
							University Pond								
Surface	A	13.20	0.2	111.84	6.869	3.67	0.188	160.67	4.109	542	1.632	402.92	0.320	117.54	0.42
	B	12.30	0	180.86	5.013	4.73	0.094	14.67	0.943	566.67	0.943	315.44	0.127	208.48	0.451
	C	8.85	0.05	148.92	4.679	5.17	0.047	146.67	0.943	593.33	0.943	329.27	0.127	107.47	0.306
Bottom	A	13.70	0.1	95.75	4.512	4.47	0.047	160.67	0.942	547.33	0.943	413.52	397	132.4	0.496
	B	13.00	0	163.14	0	4.03	0.094	16	1.633	580.67	0.943	366.97	0.116	220.41	0.502
	C	9.75	0.05	132.38	9.358	4.63	0.047	138.67	2.494	606	0	380.44	0.424	132.09	1.476

Contd...

Table 16.1(b)–Contd...

		Physico-chemical Parameters													
		Temp.		*pH*		*EC*		*TDS*		*DO*		*COD*		*PO4*	
		Mean	*SD*	*Mean*	*SD*	*Mean*	*SD*	*Mean*	*SD*	*Mean*	*SD*	*Mean*	*SD*	*Mean*	*SD*
								Katapali Pond							
Surface	A	8.85	0.05	67.03	4.512	0.63	0.047	80	2.828	364.67	0.943	396.33	2.054	171.7	0.081
	B	8.15	0.05	294.31	5.014	4.57	0.047	7.33	0.943	387.33	3.394	283.33	2.494	245.9	0.432
	C	7.15	0.05	264.73	0	4.8	0	74.67	1.886	404.67	1.886	2840	17.048	240.47	0.825
Bottom	A	9.4	0	52.27	0	0.5	0.081	86.67	942	371.33	0.943	463.33	0.471	82.56	0.205
	B	9.3	0	269.49	8.663	4.13	0.047	5.33	0.943	400.67	2.494	292.13	0.865	113.57	0.124
	C	8.15	0.05	228.33	9.358	4.23	0.047	70.67	0.943	427.33	0.943	298.33	0.471	109.57	0.518
								Golgunda Pond							
Surface	A	5.65	0.05	95.75	4.512	1.07	0.094	106.67	1.885	453.33	0.943	622.7	0.92	14.6	0.163
	B	4.85	0.15	230.49	5.014	6.77	0.047	17.33	0.943	504.67	0.943	352	0.778	213.5	0.081
	C	4.45	0.05	215.09	8.104	7.13	0.047	32	0	528.67	2.494	378.7	11.492	198.6	0.244
Bottom	A	*	*	*	*	*	*	*	*	*	*	*	*	*	*
	B	5.3	0.1	198.59	5.013	6.27	0.047	12.67	0.943	512.67	2.494	363.57	0.286	224	0.294
	C	4.8	0	182.01	12.38	6.67	0.094	29.33	0.943	542.67	3.399	391.13	0.524	211.37	0.169

A: Pre-monsoon; B: Monsoon; C: Post-monsoon
SD: Standard Deviation (Mean and SD values based on 3 replicate samples over the study period
*: Samples were not collected due to low water level of the pond
All analyses were conducted in triplicate.

Table 16.2: Average Values of the Physico-chemical Parameters [Temperature (°C), pH, Electrical Conductivity (EC mmho/cm), Total Dissolved Solid (TDS mgl^{-1}), Dissolved Oxygen (DO mgl^{-1}), Chemical Oxygen Demand (COD mgl^{-1}), Phosphate (PO_4 mgl^{-1}), Nitrate (NO_3 mgl^{-1}), Chloride (Cl mgl^{-1}), Free carbon Dioxide (CO_2 mgl^{-1}), Alkalinity (mgl^{-1}), Hardness (mgl^{-1}), Sodium (Na mgl^{-1}) and Potassium (K mgl^{-1})] of the Four Sampling Sites (Sadeipali Pond, University Pond, Katapali Pond and Golgunda Pond)

	Physico-chemical Parameters													
	Temp	*pH*	*EC*	*TDS*	*DO*	*COD*	*PO_4*	*NO_3*	*Cl*	*Free CO_2*	*Alkali-nity*	*Hard-ness*	*Na*	*K*
Sadeipali pond	31.05	8.02	0.48	0.41	5.25	81.72	460.0	3.59	92.06	3.30	45.67	499.33	56.7	213.64
University pond	30.01	7.70	1.49	0.39	3.34	106.88	1299.17	11.8	138.82	4.45	106.23	572.67	368.09	153.07
Katapali pond	29.7	7.96	0.67	0.36	7.10	69.98	1169.17	8.5	196.03	3.14	54.11	392.67	336.24	160.63
Golgunda pond	28.6	7.88	0.89	0.42	5.93	67.45	171.0	5.01	184.39	5.58	39.60	508.40	421.62	172.41

Table 16.3: Variance Values (ANOVA–Two Way of Without Replicates) of the Physico-chemical Parameters (Temperature (єC), pH, Electrical Conductivity (EC mmho/cm), Total Dissolved Solid (TDS mgl^{-1}), Dissolved Oxygen (DO mgl^{-1}), Chemical Oxygen Demand (COD mgl^{-1}), Phosphate (PO_4 mgl^{-1}), Nitrate (NO_3 mgl^{-1}), Chloride (Cl mgl^{-1}), Free carbon Dioxide (CO_2 mgl^{-1}), Alkalinity (mgl^{-1}), Hardness (mgl^{-1}), Sodium (Na mgl^{-1}) and Potassium (K mgl^{-1}) of Both Surface and Bottom in Between Ponds as well as the Seasons of the Four Sampling Sites (Sadeipali Pond, University Pond, Katapali Pond and Golgunda Pond)

	Physico-chemical Parameters													
	Temp.	*pH*	*EC*	*TDS*	*DO*	*COD*	*PO_4*	*NO_3*	*Cl*	*CO_2*	*Alkalinity*	*Hardness*	*Na*	*K*
Surface														
Ponds	2.428**	153.409*	15.979*	0.699**	4.449**	1.079**	25.100*	58.742*	4.478**	1.323**	3.090**	23.018*	21.131*	2.292**
Seasons	232.829*	18.348*	8.928*	20.429*	1.336**	4.508**	5.394*	8.472*	13.759*	11.544*	11.793*	8.530*	4.463**	4.893**
Bottom														
Ponds	5.277**	25.816*	13.913*	1.048**	12.932*	0.982**	27.682*	96.134*	3.618**	1.587**	6.033**	21.647*	40.154*	12.734*
Seasons	378.82*	2.011**	5.027**	9.917*	10.155*	2.438**	18.452*	8.298*	6.616**	2.971 **	15.777*	5.797**	1.888**	5.653**

*: Significant; **: Not Significant at F0.05.

The average phosphates content of the four ponds were 460.0, 1299.17, 1169.17 and 171.0 mg/l^{-1} in Sadeipali pond, University pond, Katapali pond and Golgunda Pond respectively (Table 16.2). So the water of University pond was relatively more in phosphate than the others. In all the ponds phosphate content in both surface and bottom water were maximum in pre-monsoon and minimum in post-monsoon except Golgunda pond. The analysis of variance (Table 16.3) showed significant difference in phosphate and nitrate, between the ponds as well as the seasons of both surface and bottom water. The chloride and alkalinity content in both surface and bottom water were maximum in monsoon and minimum in pre-monsoon. Whereas the higher and lower values of free carbon dioxide found in post monsoon and pre-monsoon respectively in both surface and bottom water of all the ponds. In all the ponds total hardness in both surface and bottom water were maximum in post-monsoon and minimum in pre-monsoon.

The average sodium content of the four ponds was 56.7, 368.09, 336.24 and 421.62 mg/l^{-1} in Sadeipali pond, University pond, Katapali pond and Golgunda pond respectively (Table 16.2). The data reveals that the sodium content in both the surface and bottom water was maximum in pre-monsoon and minimum in monsoon. The analysis of variance shows (Table 16.3) the difference in Na content to be highly significant between the ponds, where as no significance difference was found between the seasons. In general, in all the ponds the potassium contents in both surface and bottom water was maximum in monsoon and minimum in pre-monsoon. The analysis of variance shows (Table 16.3) significant difference in potassium content in between the ponds in bottom water but not the surface water, where as no significance difference between the seasons.

Discussion

During the present investigation, the surface water temperatures of ponds were lower than bottom. Maximum values of water temperature were recorded in pre-monsoon and minimum in monsoon and post-monsoon corresponding with air temperature. The low water temperature in monsoon and post-monsoon seasons can be explained on the basis of frequent clouds, high percentage of humidity, high current velocity and high water levels. But during pre-monsoon the water temperature was high due to low water level, low wind velocity, greater solar radiation and clear atmosphere (Kumar, 1999b and Singh, 1997). The water temperature is high in the pond of influx of domestic sewage (Trivedy *et al.*, 1990; Krishnamuthy and Bharati, 1997 and Singh, 1997). During the present study, it has been noted that the water temperature was comparatively higher in Sadeipali pond than other ponds.

The low pH value was noticed in monsoon and post-monsoon and high in pre-monsoon. However, a range of 7.1 to 8.7 presented a clear-cut identification of alkaline nature of pond waters. Bhatnagar and Sharma (1975) and Kumar (1996 a) found seasonal variation of pH in water. This rise in pH might be due to an increase temperature and reduces the solubility of CO_2. Here pH value was comparatively low in University pond, which might be due, to regular addition of domestic sewage.

The variance of electrical conductance was due to ionic discharges coming from household and from industrial sewage. Similar results were observed by (Nanda and Tiwari, 1999; Jayashree, 2002). However, all the four ponds showed maximum values in monsoon followed by post-monsoon and pre-monsoon. This could be associated with the lower water volume in the ponds. The surface water has maximum values than the bottom. In monsoon the values were maximum due to high input of the ionized salts. The seasonal variation in the EC is mostly due to the increased concentration of salts because of evaporation. High levels of conductivity values are useful indicators of total dissolved solids. During the present study, TDS showed certain seasonal variations. It's minimum values were recorded during post-monsoon. However, its maximum value was recorded during monsoon in all the ponds, which might be due to rainfall and influx of dissolved solid contents from the catchments area during rainfall. TDS showed variations in all the ponds. Its maximum value was observed in Golgunda pond that might be due to pouring of domestic sewage.

The dissolved oxygen (DO) is an important parameter to study the quality of water and plays a vital role in supporting aquatic life. The bottom water has lower DO. High DO content in winter is also due to the intensive photosynthetic activity on the upper water levels. The low DO content of water due to the pressure of large amount of organic pollutants resulting in increased utilization of oxygen by bacteria is also confirmed by Das (1993). In the present investigation,

DO showed distinct variation in all the ponds. The low concentration of DO in University pond might be due to the high rate of oxygen consumption by oxidisable matter coming in along with domestic sewage.

COD represents chemical oxidisable organic matter present in water hence it is a reliable parameter for judging the extent of pollution in water. The COD value is high in summer due to the decomposition process of microbes (Mahadevan and Krishnaswamy, 1988). In this study, its higher value during pre-monsoon and lower in post-monsoon indicate lower biological activity. The higher range of COD was observed in University pond that was much higher than the others. It might be due to high amount organic matter in domestic sewage.

In water, phosphorous occurs both in organic and inorganic forms (as bound in the organisimal matter), only inorganic phosphorus (as orthophosphate) plays a dynamic role in an aquatic ecosystem. The enrichment of this nutrient leads to the process of eutrophication. The increased application of fertilizers, use of detergents and domestic sewage play a great role in contributing the heavy loading of phosphorous in water (Ghosh and Sharma, 1989) also reported increase of phosphate as a result of sewage contamination. The university pond has comparatively maximum values than the others due to (I) washing the phosphate containing rocks present in near hills, (II) from detergent used in bathing and (III) from the agricultural runoff. The reduced level of phosphate in winter might be due to biological removal. The values of phosphate were also observed fluctuating from season to season. Its values were maximum during pre-monsoon possibly due to high evaporation. The solar radiations have encouraged the biological degradation of the organic matter and subsequent release of more phosphate to the aquatic system.

In some cases, the maximum values of nitrates were recorded during rainy season, which might be due to influx of nitrogen rich floodwater that brings large amounts of contaminated sewage water (Sunder, 1988). However, the nitrate values were sufficiently found higher in summer season too, which clearly shows the pollution load of pond. These higher values in pre-monsoon could be due to (I) decomposition of organic matter, which gets transferred by aerobic and anaerobic bacteria at high temperature into nitrogen organic matter (II) addition of NO_3-N from near by runoff. Its value was higher in University pond than the others.

Table 16.4: Ranking of Water Bodies in Order of Extent of Pollution

Parameter	*Pond*			
	Sadeipali Pond	*University Pond*	*Katapali Pond*	*Golgunda Pond*
Temperature	****	***	**	*
pH	*	****	**	***
EC	*	****	**	***
TDS	***	**	““	****
DO	***	****	*	**
COD	***	****	**	*
Phosphate	**	****	***	*
Nitrate	*	****	***	**
Chloride	*	**	****	***
Free CO_2	**	***	*	****
Alkalinity	**	****	***	*
Hardness	**	****	*	***
Sodium	*	***	**	****
Potassium	****	*	**	***
Total	**30**	**46**	**29**	**35**

*: Pollution load.

The high concentrations of chloride are indicators of large amount of organic matter in the water Bhatnagar and Sharma (1975) and Adoni (1985). Munuwar (1970) suggested that higher concentrations of chloride in water are an index of pollution of animal origin and there is a direct correlation between chloride concentration and pollution level. Minimum values of chloride were recorded during pre-monsoon and maximum in monsoon. This may be certainly due to influx of floodwater rich in chloride. The higher values of chloride were recorded in Katapali pond, which may be due to continuous influx of chloride contaminated domestic sewage. During pre-monsoon, the free CO_2 was observed in its maximum values in University pond. This can be attributed to the accelerated rate of decomposition of organic matter by microbes, decreased photosynthetic activity and high rate of respiration by benthos and microbes due to the influx of sewage. The bottom water shows the higher alkalinity than the surface during pre-monsoon. Where as the bottom water have lower value in monsoon and post-monsoon. This may be due to low water level and lower temperature bringing down the rate of decomposition of salts to a minimum there by increasing the alkalinity. University pond has high alkalinity and hardness, which might be due to the domestic sewage.

Table 16.5: Classification and Zoning of Water Bodies (CPCB, 1979)

Sl.No.	*Designated-Best-Use*	*Nomenclature for the Class of Water*
1.	Drinking water source without conventional treatment but after disinfection	Class A
2.	Outdoor bathing	Class B
3.	Drinking water source with conventional treatment followed by disinfection	Class C
4.	Propagation of wildlife, fisheries	Class D
5.	Irrigation, industrial cooling and controlled waste disposal	Class E

Table 16.6: Class of Water*

Sl.No.	*Parameter*	*A*	*B*	*C*	*D*	*E*
1.	pH	6.5–8.5	6.5–8.5	6.5–8.5	6.5–8.5	6.0–8.0
2.	EC (mmho/cm)	–	–	–	1.0	2.25
3.	TDS (mg/l)	500	–	1500	–	2100
4.	DO (mg/l)	6	5	4	4	
5.	Nitrate (mg/l)	20	–	50	–	
6.	Chloride (mg/l)	250	–	600	–	600

Sodium and potassium are highly soluble in water and impart softness to water. These are said to be utilized by blue green algae in high concentration (Provasali, 1969). High concentration of sodium in the water affects the physical conditions of soil (Rawal, 1978). High potassium level found in University pond where as high sodium in Golgunda pond, due to the geo-chemical strata containing lateritic surface soil.

Table 16.7: The Minimum and Maximum Value of Each Parameter

Parameter	Pond							
	Sadeipali Pond		University Pond		Katapali Pond		Golgunda Pond	
	Min	Max	Min	Max	Min	Max	Min	Max
Temp (°C)	25.2 ±0.081	38.8 ±0.081	24.3 ±0.047	37.4 ±0.047	24.9 ±0.047	35.5 ±0.047	24.6 ±0.081	35.8 ±0.047
pH	7.94 ±0.004	8.13 ±0.008	7.63 ±0.008	7.67 ±0.004	7.90 ±0.004	8.09 ±0.004	7.72 ±0.009	7.95 ±0.003
EC (mmho/cm)	0.32 ±0.001	0.64 ±0.002	0.86 ±0.001	2.02 ±0.002	0.46 ±0.001	0.83 ±0.003	0.53 ±0.001	1.21 ±0.001
TDS (mg/l^{-1})	0.13 ±0.001	0.67 ±0.001	0.30 ±0.003	0.52 ±0.001	0.20 ±0.009	0.49 ±0.001	0.22 ±0.001	0.55 ±0.001
DO (mg/l^{-1})	2.38 ±0.245	7.47 ±0.075	2.00 ±0.226	4.96 ±0.569	5.87 ±0.150	8.21 ±0.272	3.79 ±0.150	8.27 ±0.199
COD (mg/l^{-1})	28.07 ±1.819	163.99 ±47.37	47.97 ±1.886	166.65 ±4.989	39.99 ±1.886	89.32 ±0.0	23.97 ±3.771	98.65 ±3.771
Phosphate (mg/l^{-1})	90 ±40	760 ±0	925 ±25	1735 ±65	715 ±45	1950 ±20	90 ±40	255 ±125
Nitrate (mg/l^{-1})	2.41 ±0.01	4.40 ±0.0	8.85 ±0.05	13.70 ±0.10	7.15 ±0.05	9.40 ±0.0	4.45 ±0.05	5.65 ±0.05
Chloride (mg/l^{-1})	31.94 ±4.512	148.96 ±5.013	95.75 ±4.512	180.86 ±5.013	52.27 ±0.0	294.31 ±5.014	95.75 ±4.512	230.49 ±5.014
Free CO_2 (mg/l^{-1})	1.57 ±0.047	4.67 ±0.094	3.67 ±0.188	5.17 ±0.047	0.50 ±0.081	4.80 ±0.0	1.07 ±0.094	7.13 ±0.047
Alkalinity (mg/l^{-1})	5.33 ±0.943	76.00 ±1.632	14.67 ±0.943	160.67 ±0.942	5.33 ±0.943	86.67 ±1.886	12.67 ±0.943	106.67 ±3.399
Hardness) (mg/l^{-1})	406.67 ±3.399	560.00 ±1.633	542.00 ±1.632	606.00 ±0.000	364.67 ±0.943	404.67 ±1.886	453.33 ±0.943	542.67 ±3.399
Sodium (mg/l^{-1})	48.0 ±1.202	65.43 ±0.262	315.44 ±0.127	413.52 ±0.397	283.33 ±2.494	463.33 ±0.471	352.00 ±0.778	622.70 ±0.920
Potassium (mg/l^{-1})	156.80 ±1.067	261. 60 ±0.355	107.47 ±0.306	220.41 ±0.502	82.56 ±0.205	245.90 ±0.432	14.60 ±0.163	224.00 ±0.294

Conclusion

When the values of different physico-chemical parameters were scored from 1-4 star marks (*) in order of pollution (Table 16.4), the University pond got the highest score of 46 followed by Golgunda pond and minimum pollution was observed in case of Sadeipali pond and Katapali pond. The Central Pollution Control Board, New Delhi (CPCB, 1979) has classified the surface water based on best-designated use (Table 16.5). Table 16.7 gives a comparative picture of the quality parameters in four ponds. Based on the values prescribed in Table 16.2, it is observed that all the ponds do not satisfy the requirement for use under A, B and C in respect to dissolved oxygen indicating a high organic load. Although study of some more parameters including the microbial status was desirable, due to time constraints, all the parameters could not be analyzed to assess the overall quality of the water bodies.

Acknowledgement

Author expresses his gratitude to the Head, Department of Environmental Sciences, Sambalpur University Orissa for providing necessary laboratory facilities.

References

Adoni AD (1985) Work book on limnology, Dept. of Botany, Dr. Harisingh Gour Vishwavidyalaya, Sagar, 210 P.

Batcha ASM (1998) Studies on hydrography and domestic pollution problems in the North bank of river Cauvery. J. Env. & Poll. 5: 69-71.

Bhargasa DS (1983) Environmental Pollution (Series B) 6:51-57.

Bhatnagar GP and Sharma GP (1975) Physico-chemical features of sewage polluted lower lake, Bhopal. "Environments and their biological effect" Proc. of the Inter. Symp.Int. 5: 212 -223.

CPCB (1979) Central Pollution Control Board 1979.

Das NK and Sinha RK (1993) Organic waste and its effect on macro-zoobenthos in Ganga at Patna (Bihar), India J. Freshwater Biol. 5: 33-40.

Ghosh NC and Sharma CB (1989) Pollution of Ganga river. Ashish Publ. House, New Delhi. PP. 246.

Goel PK (1997) Water pollution: Causes, effects and control. New Age International (P) Limited Publishers. P. 269.

Hague I Bardoloi, *et al.* (2002) Physico-chemical parameter of Toklai river, Assam. Indian J. Env. Proto 22(2): 151-153.

Jayashree J (2002) Quality of water in Parvathy Puthanar in Thiruvanathapuram. Indian J. Env. Proto 22(2): PP. 188-190.

Jha AK, Latif A and Singh JP (1997) River pollution in India: An overview. J. Environ. Poll. 4:143-151.

Kandhasamy M and Santhaguru K (1994) Influence of sewage on Physico-chemical characteristics of the river Vaigai. J. Ecobiol. 6: 315-317.

Krishnamurthy SR and Bharati SG (1997) A study on concentration of chloride of the river Kali near Dandeli, Karnataka (India). J. Environ. Pollu. 4: 9-15.

Kumar A (1996 a) Impact of organic pollution on macro zoobenthos of river Mayurakshi in South Bihar. Poll. Res 15: 85-88.

Kumar A (1996 b) Comparative studies on diel variation of abiotic factors in lentic and lotic freshwater ecosystems of Santa I Parganas (Bihar), India. J. Environ. 3: 53-56.

Kumar A (2002) Biomonitoring of Sewage pollution, APH Publishing Corporation, New Delhi, 1-402 P.

Mahadevan A and Krishnaswamy S (1988) Ma and the River Vaigu. In: Trivedy. R.K. (ed). Ecology and Pollution of Indian Rivers APH, New Delhi, India: 361-378.

Mukherjee B (1997) Environmental Biology, Tata McGraw-Hill Publ. Co. Ltd., New Delhi, 691 P.

Munawar M (1970) Limnological studies on freshwater ponds of Hyderabad India I. The Biotype. Hydrobiologia. 35: 127-162.

Nair J and Ganapati S (1997) Water quality of the Bhadar river basin. Ind. Environ. Hlth. 39: 197-206.

Nanda SN and Tiwari TN (1999) Effect of discharge of Industrial effluents on the quality of river Brahmani at Rourkela. Indian J. Env. Proto 19(1): 52-55.

Pervez S and Pandey GC (1994) Contamination of river water and sediments by thermal power ash pond discharge. Ind.J.Environ. Hlth. 36: 8-12.

Pinckney H *et al.* (2001) The Role of Nutrient Loading and Eutrophication in Estuarine Ecology. Environ. Health Perspt. Vol. 109 (5).

Prajapati R and Mathur R (2002) Ground water quality in Sheopurkalan town. Indian J. Env. Prot. 22(2): 197-200.

Provasoli L (1969) Algal nutrition and eutrophication, In Eutrophication: Causes, Consequences, Correctives, Natural Academy of sciences, Washington D.C: 574-593. Pure and App1. Limnology (Ed. A.D. Adoni). Bull. Bot. Soc., Sagar. 32:

Rawal NC (1978) Quality of river water of India, Proceedings of 47th Research Session of the CSIB, Hubli Dharwad, Karnataka, Vol.-2: 139-160.

Sahu BK, Rao RJ and Behera SK (1994) Studies of some physico-chemical characteristics of Ganga river water (Rishikesh-Kanpur) within twenty four hours during winter, 1994. Ecology, Environment and Conservation. Vol.1.

Sahu BK, Rao RJ, Behera SK and Pandit RK (1995) Diel variation on some abiotic parameters of the Ganga river water (Rishikesh-Kanpur) in mouth of August. J. Ecotoxicol. Environ. Monit. 5: 119-124.

Saxena KK and Chauhan RRS (1993) Physico-chemical aspects of pollution in river Yamuna at Agra. Poll.Res 12: 10 1-104.

Singh AK (1997) Limnobiotic studies of river Mayurakshi in Santal Parganas, Bihar. Ph.D. Thesis, S.K. University, Dumka.

Singh NK, Kumar A, Ahmed MF and Yadav NK (1982) Ground water contamination with special refence to fluoride and iron ions in some area of Bhagalpur. J.Freshwater Biol. 4: 159-162.

Sinha KK and Kumar A (1983) Ecological management of Industrial effluents of soft drink bottling factory in Jaipur (India). Biol. 5: 21-28.

Sunder S (1988) Mounting the water quality in a stretch of river Jhelum, Kasmir. In: Ecology and Pollution of Indian Rivers. (Ed. R.K. Trivedy). Ashish Pub1. House, New Delhi, pp. 131-161.

Trivedy RK, Shrotri AC and Khatavkar SD (1990) Physico-chemical characteristics and phytoplankton of the river Panchganga near Kolhapur, Maharastra. In. River Pollution in India (Ed. R.K. Trivedy). Ashish Publishing House, New Delhi. pp. 159-178.

Author Index

Subject Index